Rocket Surgeon

By

A.E. Williams

DISCLAIMER:

The *'Terminal Reset* Series' novels are works of **fiction** and deal with **fictional** events. Even if some locations depicted do exist and some collective events did occur, this story is entirely fictitious. Most of the characters therein are a figment of the author's imagination. Without exception, those characters that are historical figures of fact or based upon historical figures of fact are used fictitiously, and their actions, demeanor, conversations, and characters are similarly all figments of the author's imagination. Any resemblance to reality would be a coincidence.

Published by

www.literarysherpa.com

Copyright © 2017 by **Literary Sherpa, LLC,**

and A.E. Williams

All rights reserved. No part of this publication may be reproduced, distributed or transmitted in any form or by any means, including photocopying, recording or other electronic or mechanical means without the full prior and written permission of the publisher. Exceptions include brief quotations for review and certain other non-commercial uses as permitted by copyright law.

For permission requests, contact the publisher at:

Literary Sherpa, LLC

Contents
- Foreword 1
- Introduction 25
  - Science Degree – First Iteration 34
  - Science Degree – Second Iteration 41
  - HR Degree – First Iteration / Science Degree – Third Iteration 44
  - Science Degree – Fourth Iteration 47
- Speculative Fiction Showcase Articles 62
  - Why MANNED Space Exploration? A Treatise on the Necessity for Manned Spaceflight 63
  - MANNED Space Exploration? It's Getting Old – PART 1 80
  - MANNED Space Exploration? It's Getting Old – PART 2 97
  - How Spaceflight and The Challenges Therein Have Been Addressed in Science Fiction 117
  - Curse You, Albert Einstein! 131
  - Why the Movie Version of "The Martian" Isn't About Mars – Or Science 149
  - Coming Soon! 162
- Apocryphal Tales from the Vault and Other Random Thoughts 164
  - Rocket Surgeon 164

- Relativity in Action 206
- The Scientific Method 207
- The Parable of the Ants 228
- Jayvonne the Steely-eyed 237

Random-ish Thoughts 242
- On Artificial Intelligence, The Federation, and Utopia – or I Have No Mouth, and I Must Concatenate! 242
- Militarized Drones and You – or Why "Terminator" Wasn't Supposed to Be an Instruction Manual .. 255

Closing Thoughts 260
- Our Future Is in the Stars 260
- Happiness IS a Choice 262

SPECIAL BONUS EXCERPT 267
TERMINAL RESET – THE COMING OF THE WAVE ... 267
- PROLOGUE 267
- CHAPTER ONE 277

Credits 296
About the Author 296

Foreword

"It has been my observation that the happiest of people, the vibrant doers of the world, are almost always those who are using - who are putting into play, calling upon, depending upon-the greatest number of their God-given talents and capabilities." -- John Glenn[1]

"Some dreamers demand that scientists only discover things that can be used for good. That is impossible. Science gives us a powerful vocabulary, and it is impossible to produce a vocabulary with which one can only say nice things." -- John Polanyi

"We're all dreamers." -- Ray Bradbury

---

[1] This work was in progress when we received news of John Glenn's passing. He was a true American hero, and one of the privileged few to see our planet from the vantage of an orbiting spacecraft. He will be sorely missed. *Ad Astra*, and Godspeed, John Glenn.

A.E.Williams

✲ ✲ ✲ ✲ ✲

There are two kinds of people in this world - Dreamers and Doers.

Into this world come those to whom it is evident that Science and Science Fiction are two sides of the same coin - one can't exist without the other.

The Doers cannot act on anything unless the Dreamers first conceive of the seminal concept; the Dreamers rely on Doers to give birth to their dreams.

It is thought by some that most Dreamers don't understand the physics or the engineering principles behind the reality of Science.

Perhaps it is this lack of understanding that allows them their flights of fancy, their follies, and their impossible dreams?

A vivid example of this is the various Space Programs of our planet.

As of today, the engineers have brought forth a particular vision, a reality bound tightly and irrevocably to our current scientific knowledge.

Rocket Surgeon

Logic and the hard edge of mathematics are the hammer and anvil of our progress.

But then, look at the Space Programs of our world as the Dreamers have imagined it.

The technology and wonders envisioned in Science Fiction present us with a diametrically opposed viewpoint.

Unfettered by the necessity of logic, math and physics that hinder our Reality, they posit methods by which we compress space and time, in order to better serve our needs to explore, spread out, and delve into the mysteries of this Universe.

One wonders what the shepherds and seamen of yore thought when they saw the moon in the sky?

Five-hundred years isn't that long a time, but it is enough to separate their beliefs and superstitions from ours.

Could we suppose that they may have dreamt of walking on it?

A.E.Williams

How would they have posed the problem?

Did they think of flying to it?

Or is this just something that has preoccupied Mankind in the past 100 to 125 years?

I personally think it's because mankind has always dreamed of the miracle of Flight.

He dreamed of flying to the stars, or flying somewhere. Birds aloft were magical creatures, to be envied their freedom from the bonds of gravity (then an unknown concept, we suppose).

We've known birds can fly for hundreds of years; perhaps Man wondered if he could mimic them?

After all, he celebrated Flight.

He created art about it.

Leonardo Da Vinci drew sketches of the wings of birds, of their anatomy.

We wrote songs about it, and created legends and myths around it.

And, we observed them.

We studied them.

Rocket Surgeon

We hoped to finally be them.

And so, man eventually became like the birds, albeit with some mechanical assistance.

The engineering principles and understanding of the physics behind the Flight took thousands of years to be developed into a working understanding of it – the Science of it.

I believe that it'll probably take hundreds of years more in order for mankind to understand the physics and engineering principles involved and traveling to the stars, safely and efficiently.

But, I am impatient.

❋ ❋ ❋ ❋ ❋

Did you realize that original series of the television program *"Star Trek"*, first televised in the 1960's, was merely a stylized version of "Wagon Train"?

Set somewhere in a highly stylized and imaginative future, each episode was self-contained and presented morality crisis that the crew of a spaceship had to overcome.

It was a microcosm of our own troubled world.

The viewer hopefully would grow, just a bit, and come away feeling better about themselves, learning how to live with someone else or overcoming prejudices.

It was an interesting experiment, showcasing technological marvels that staggered the imagination.

"*Star Trek*" also fired the imagination of a generation of young people, and made them desire that kind of world.

It gave birth to a tremendous growth in the interest of Science, Technology, Engineering and Math (STEM) degrees.

Its impact was felt over the next 30 to 40 years, as the fictional Universe of Star Trek was more and more often integrated in our own Universe, in which the advances in engineering and of course, social awareness brought us closer to that vision.

Let's take, for example the ubiquity of cell phones, 3D printers, and microwave ovens.

These were all concepts or visions that were put forth as Gene Roddenberry's attempt to explain the future.

They were the props and stage decorations that

lent his vision of that future a kind of verisimilitude, an attempt to make real the imagined.

Fifty years later, after its debut, we notice how the engineers took those concepts and brought them to reality.

The Dreamers thought about it and the Doers made it real.

Think of what the next 50 years of engineering and physics advancements will bring![2]

Also, when I think of Dreamers, I think of Roddenberry's hope of a positive-type future.

The world he created was only marginally dystopic, for effect.[3]

It is my sincere hope that the Doers of today remain steadfast in their ability to turn Dreams into

---

[2] Can we say "transparent aluminum " please? Oh, wait, they got that too.

[3] A lot of authors that wrote novels in the seventies, eighties and nineties about apocalyptic ends of the world, yet here we are in 2017, almost five years after the supposed end of the Mayan calendar, and we're still hale and hearty. The election of 2016 may have changed that, of course!

Reality, and the authors of speculative fiction provide the grist for the mills of the Dreamers, so they may never run out of the impossible ones.

✳ ✳ ✳ ✳ ✳

"Where do you get your ideas from?" is probably the second most asked question authors get.[4]

Since we started writing the smash science fiction series *"Terminal Reset"* in late 2013, readers have asked how the concept came about.

What does it have to do with all this prior rant regarding science and science fiction, all this Doer and Dreamer gobbledygook?

To answer that question, it's necessary to have an **Annoying Autobiographical Pause**.

✳ ✳ ✳ ✳ ✳

---

[4] "Where is the bar?" is most likely the first.

Rocket Surgeon

Over the last few years, I had been thinking about my life.

Had it been as complete as I envisioned it when I was younger?

Was there anything that I wanted to do over again?

Did I have regrets?

Of course, most people my age (50-plus) have these thoughts - daily.

Normal people kind of dream or fantasize about the road not taken, what might have happened at key moments when life took a fork in the path.

Everyone has something in their life that they would like to do over again.[5]

I've seen time travel books that attempt to rectify this by sending a version of the current Wiser You™ back to try to knock some sense into stupid, younger HUYOA[6] You™.

But, seriously, who wants to go back in time at

---

[5] Hell, the entire game of golf is based on mulligans! Or, at least, rules that favor mistakes.

[6] Head-Up-Your-Own-Ass™

their current age just to see HUYOA You™ fuck up, in spite of Wiser You™'s sage advice and even advance warning of their mistakes?

Besides, that's been done ad nauseam, and I wanted to explore the idea of second chances in an entirely different manner.

There was a book I read about 20 or 30 years ago about a guy who dies, and then wakes up in his body of 25 years prior, but with all his memories intact[7].

The concept was sound, and entertaining, but I thought that kind of boring to do that when it's only yourself, or maybe another one or two others that are affected.

It sounds more like an old man rant, than something exciting and fresh.

But, I liked the idea of -*wisdom*- that was baked into the theme.

❋ ❋ ❋ ❋ ❋ ❋

---

[7] The awesome *"REPLAY", by* Ken Grimwood

Rocket Surgeon

Since I received my Kindle, I've read voraciously.

I devour all manner of science fiction, alternate history, and other genres. I feel grateful that I can see the thoughts of other independent authors, and even admire many of their successes.[8]

I noticed a tremendous amount of end-of-world (both apocalyptic and post-apocalyptic) books and movies. I've always enjoyed these, and apparently, many others share my taste.

And, being a bit of a writer myself, I decided to try to marry the concepts of time travel, replays, and the correction of past mistakes, and the idea of retained memory.

But, I still needed a vehicle to carry my nascent imaginings.

---

[8] Hugh Howey, of course, but also Wayne Stinnett and Ryk Brown provide me with rollicking good fun!

The final pieces came about four years ago, while, when watching a *"Star Trek"* marathon, I viewed both *"Star Trek VI"* and *"VII"*, and then *"Armageddon"*.

I thought about all the pieces of my personal puzzle, floating around, just waiting to be assembled.

There were the psychological elements of my initial concern –

- Mortality
- Regret
- Do-overs
- Redemption
- Evolution

Maybe even Salvation?

There were the parameters of how these things connected –

- Memory
- Human Relationships
- The Human Condition

I pondered the speculative and unknown Universe that might exist if some outside force, an asteroid, or something else, created conditions that would allow me to explore these ideas further.

Rocket Surgeon

I asked myself this question:

"What if a wave of energy[9] come out of deep space, but instead of a force of destruction like a comet or asteroid or even the energy itself, it would mysteriously de-age (regress) all life over the entire planet? This radiation would have no effect on the inert components of the planet itself. Rocks, water, air, and other things that allow our existence wouldn't be affected. But, organic, living creatures would."

How could I create a situation where this might help to find a more comprehensive understanding of those ideas I was contemplating?

As I continued to think about it, I began to write down some of these thoughts.

I played around with it for a few days, thinking about the concepts, clarifying the rules for this world.

---

[9] One of the fun things about writing *'Terminal Reset'* has been the sheer number of interesting coincidences that have arisen based on our thorough research of topics. For instance, this article talks about a cloud of gas on a collision course with the Milky Way. http://www.businessinsider.com/nasa-mysterious-cloud-collides-galaxy-2016-10

I don't even remember when it hit me that I had enough for a book, but I didn't really seriously consider writing one at that point.

However, my research quickly came up against a blank Google search screen.

I wondered, had no one written a story where this been even thought of before?

This seemed unlikely, but also gave me a great opportunity.

Later, I started to think, about the logistics as to how could I make the story believable.

I examined how good authors can you tell a story that resonates to your core.

There are techniques they use that allow your emotions, your feelings, to be brought to the surface.

The technical details in the background are important, to be sure, for science fiction.

Especially hard sci-fi.

By bringing some of my concepts together (not all, remember, I'm still a dreamer at this point) I sought to begin to lay out the skeleton of this creature.

As I mentioned, a marketable work would have to deal with the current genre focus, and that was APOC Sci-fi.

So, the first hurdle to be cleared had to be how could we do this aging effect realistically,[10] and bring it into the realm of apocalyptic science fiction?

This begat the question as to just how old people should be NOW, in order to cause the de-aging to be significant.

Was the degree of de-aging (or *Regression*, as it came to be known in the book) only to be a ten-year span?

So, a person of 25 would end up at 15 years after *Regression*?

How about twenty years?

Fifty years?

Maybe one hundred years would work?

---

[10] Oddly enough, realism is the key component to good storytelling, especially regarding fantastic tales!

After playing around a bit, I came up with some rudimentary algorithm that made a kind of logical sense.

I couldn't use a hundred years for a *Regression* effect, since there wouldn't be anybody left but a handful of people.

A ten-year *Regression* effect cause enough of a societal change to matter to the storyline.

I decided to make the effect of *Regression* subtract 40 years.

This would cause an approximate population decrease in the world population of about 60%

It was enough of a reduction to the surplus to achieve our goal of Apocalypse, but not enough to endanger the continuation of the human race.

The second major hurdle was the composition and rationale behind *The Wave*.

Just what is this mysterious radiation?

How was it formed?

Why?

It was some exposure to the modern theories of

Quantum Mechanics and physics that provided the superficial answers to these questions.

As I researched more and more the workings of our world to provide the rules engine for *"Terminal Reset"*, these questions became answered.

But, there loomed a much larger one.

In spite of all of our scientific achievements and exploration of space, and our probes and computer, we are still VERY uncertain of just how certain elements of the Universe work.

We have theories, and ideas.

Many of these allow us to create the dreams of which I have alluded earlier:

- Communications devices that fit in our palms.
- Celestial navigation for everyone, for a pittance[11]
- Powerful computers that answer almost anything we can think of to question
- Medical miracles

We also are staggeringly ignorant of many things:

- How fire works
- Why we sleep when we do
- Fractals in nature, and why these particular patterns can be found under enormously variant conditions, yet appear identical
- The forward-only nature of time
- The propagation of radiation in space

One of the beautiful things about fiction, (and especially *speculative* fiction!), is that we don't have to explain in detail the 'real' science of why things happen.[12]

---

[11] No more need of compass and sextant, or an understanding of geometry!

[12] Warp drive, wormholes, teleportation and other artistic

But, the best science fiction at least attempts to take a stab at **defining** the rules of its own fictional universe, so as to at least give the glimmer of consistency.

So, in thinking about *The Wave*, and what were the effects that it would have on the planet, it was decided that it would have an effect only upon *organic* life.

All organic life whether it be animal, or vegetable.

I decided that water could act as a mitigating agent to this radiation, because science has proven that water is a good insulation for the majority of types of radiation with which we have experience.[13]

These few rules became the framework of the foundational concepts that grew into the world of "Terminal Reset".

---

conveniences can be 'hand-waved' and the reader accepts these conceits to be able to enjoy the story.

[13] Yes, yes... I know that the particles that become suspended in water are STILL radioactive, and form isotopic compounds that will remain radioactive until they decay. But, the alpha, beta, and gamma rays do lose strength with distance, and the density of H2O guarantees some decay due to incipient impacts at the molecular level.

A.E.Williams

Rocket Surgeon

✵ ✵ ✵ ✵ ✵

Reader feedback had fallen into some predictable categories, but I felt it important that they remain.

The setting of the story in our immediate world was not incidental, and provides for an alternate-history parallel Universe, where things are immediately familiar and relatable.

Some of the fictional characters have real-world analogs, but mostly as a convenience to allow the reader to identify the workings of governments.[14]

The addition of some *creative* plot elements has raised eyebrows among some readers, but there is enough literary historical precedent[15] to allow for the inclusion of these types of things in science fiction writing.

It is hoped that the entire fabric will be seen for what it is, and not that a focus on some of the wilder and more fantastic pieces distract from the

---

[14] Plus, it's kind of fun to pretend that certain people might act or react in certain ways!

[15] I'm looking at you, *"League of Extraordinary Gentlemen"*, *"Stargate"*, and *"Prometheus"*!

central ideas.

I hope you have enjoyed this peek behind the curtains as to how an author might come up with an idea.

I tried to give a bit of background about the process that was hashed out, prior to launching *"Terminal Reset"*.

Of course, there are many more concepts that still have to be worked out in the years to follow.

We plan to follow in the footsteps of Gene Roddenberry and bring you a brand-new Universe to explore, hopefully packed with the same kind of insight and challenging ideas that were the hallmark of *"Star Trek"*. [16]

---

[16] But, this is NOT *"Star Trek"*, is it? One of the challenges authors have is the idea of derivative works. Everyone in the world has ideas. Everyone has dreams. And, to be fair, there is a good deal of 'sharing' of these ideas amongst the artistic works of our world. While Roddenberry himself was innovative, it was *"Wagon Train"* that he emulated.
And, *"Wagon Train"* was really just the televised version of old Western films, whose pedigree is as old as Homer's *"Odyssey"* and *"Iliad"*. The Hero's Journey, best recently exemplified by *"Star Wars"*, is not the exact same trip as *"Star Trek"*. In the former, we see the growth of a person who is ignorant, and who comes to their complete self-actualization by dint of luck and divine intervention. (Yoda is essentially a god, and

Rocket Surgeon

Enjoy the ride.

A.E. Williams

High Springs, FL

November, 2016

Obi Wan is Jesus himself!) The world of *"Star Trek"* presents itself as a series of adventures, but the main idea to pull away is that of the hero questioning the rationality of the situation as filtered by Logic and Emotion. Kirk is the Human in all of us, trying to make some sense of the world. Spock is the rational side of Argument (and not totally). McCoy is the Emotional and Feeling side. But, this is basically a battle between the Ego and the Id, as old as Shakespeare's *"The Tempest"*, or it's more recent incantation, *"Forbidden Planet"*. The genius of Roddenberry was being able to engage us at the most entertaining level to consider what effect we, as people, were having on the Human Condition.

Introduction

One night in 1966, I was allowed to 'stay up' and watch a startling new television show. It was called *"Star Trek"*, and this show's episode was called *"The Mantrap"*.

At the tender age of five, my experience with shows were limited to Captain Kangaroo, and Saturday morning cartoons, such as "Jonny Quest", and "Space Ghost".[17]

It was a much simpler time, full of hubris and American Pride.

The recent death of John F. Kennedy had not taken root and morphed into the cynicism and hypocrisy that was on the horizon.

---

[17] You may be surprised to find out that *Space Ghost* didn't have a Coast-to-Coast television show of his own back then. No, he just flew around in the Phantom Cruiser, blasting lava monsters, flying from place to place like Superman, and shooting rays from his wristbands. He was super-cool, and my first real Halloween costume was made from scratch to emulate him. Unfortunately, the needed materials to create actual ray-bands did not exist at the time. It was the first of many disappointments that science fiction did not really mirror reality. Yet.

Most television shows were full of white men, doing empire-building activities, with little regard for consequences.

The white cops shot white criminals, and seemed to have white wives and white girlfriends.

The white cowboys shot suspiciously Mexican-looking Injuns,[18] who rode around threatening to scalp and take the white women.

This backdrop of unreality gave an impressionable young boy dreams of riding horses, shooting people who got in his way, and riding off into the sunset with some presumably happy wimmen-folk tending the home fires, and cooking grub, when he would triumphantly return from whatever adventures he found out on the range.

I can tell you quite frankly that this naïve view of things has radically changed over my lifetime.

There have been periods in my life of necessary adaptation to the reality of the world, as contrasted by the morality tales and fables that were thrust upon me over the decades.

---

[18] Hey, that's what we called them!

Rocket Surgeon

But, the first inkling of these changes came in the form of a space-western.

The men were still 'real' men, (even though Kirk was played by *Canadian* William Shatner), and the women were pretty and lithe and – black?

Well, that wasn't anything odd.

I mean, I grew up down South.

Plenty of Negroes where I lived.[19]

But, there were also a Jap and a Ruskie on the show.

And some Scottish dude.

And an old-timey Southern doctor.

And that hot nurse...

And that Devil-looking guy, only he was green, not red.[20]

---

[19] Again, that's what we called African Americans in the 1960's...and the Reader should keep in mind that is NOT a politically correct term. It is used here for effect, not to race-bait anyone.

[20] On my 9-inch Cathode Ray Tube television set, these people were various shades of grey.

They all traveled around space – sorry, 'trekked' around the stars – in the weirdest looking space ship anyone had ever seen.

It wasn't 'exactly' a flying saucer.

It had wings with tubes, and some kind of hull beneath the saucer part.

Weird.

And then, the story began for real, right after the credits rolled, and I was sucked in by the salt-vampire that killed the men.

To say that I found it terrifying is a tremendous understatement.

I think I probably shit myself.

The final scene, when Nancy transformed into the most hideous monster ever seen on the small screen[21] left me shaken for YEARS.

But it also lit the kindling that blazed into the passion for science and speculative fiction that I have enjoyed throughout my life.

---

[21] Until the recent Presidential elections...

Rocket Surgeon

I can't say *"Star Trek"* was responsible for me becoming a scientist, engineer, and computer genius for sure, but it certainly was a golden thread woven into that fabric.

My love of that feeling of tension and suspense, buoyed by the hope that we'd soon get flying cars and jetpacks[22] dovetailed conveniently with the *real* space race occurring literally in my own back yard.

I lived in Florida at the time, and remember seeing the Apollo 11 rising into the blue Summer skies, from behind the windows where I was cloistered with all the other students and the nuns of the Catholic school my parents had kindly chosen for my education.

I flashed back to that scene many years later, when I was readying to go to work at the aerospace company where I was employed, looking in horror as Challenger destroyed itself – the angles of the launch where near-identical, as they used the same complex[23].

---

[22] See *"The Jetsons"* and *"Thunderball"*

[23] Apollo 11 launched from Pad 39-A, but Challenger launched from Pad 39-B. Both are part of the Kennedy Space Center

This juxtaposition of the greatest technical rocketry achievement of the Twentieth Century with one of the two most disastrous events[24] in modern space science I see as pivotal in determining my opinions regarding manned space flight.

During my formative years, the Cold War and Space Race were in full throttle mode.

On the one side, we were ducking and covering under our desks, hoping to avoid nuclear annihilation for another millisecond.

On the other, we were reaching for the stars, trying to expand beyond the parochial attitudes and prejudices of our tiny world.

When I reached the age where I was able to read fluently, I devoured both science and fiction stories.

Oh, of course we had to read books in school.

Everyone got exposed to those morality tales such as *"Alas, Babylon"* or *"Slaughterhouse Five"*.

Launch Complex 39. Source: https://en.wikipedia.org/wiki/Kennedy_Space_Center_Launch_Complex_39

[24] *Challenger* and *Columbia*

Rocket Surgeon

Some of the more enlightened schools might throw fantasy a bone with *"The Hobbit"*, or *"Stranger in a Strange Land"*.

The more cultured schools would force *"Wuthering Heights"* and *"Jane Eyre"* on you, as well as Shakespeare's more obscure works.

To provide some illusory semblance of balance, you'd probably get *"The Jungle"*, *"The Pigman"*, or an anthology of Poe's tales of horror.

In other words, kids those days were reading the traditional muck that passed for great literature; stories that threw the previous centuries into sharp relief; the 'classics' of ancient Rome and Greece; the Eurocentric tripe that elevated dead monarchies to exalted levels, that were to be emulated by we unwashed masses.

These words strove hard to forever cement the idea of 'one's bettors' in young, impressionable minds. [25]

The heroes of the day were classic in the sense that

---

[25] To this day, I feel that our society performed a great disservice in that entire generations of our youth were manipulated in this way. We fought against monarchy, and yet still seem to put it on a pedestal as the highest form of government.

the gods had favored them, and their kings, with divine rights.

They were not to be questioned.

The lottery of birth had been won and claimed by them, and you were lucky to breathe the same rarified air as they, even vicariously.

So, yes, I read all this crap, and wrote essays on ideas and words that were 'revolutionary'.

A century of elite towers of Academia couldn't be wrong, could they?

But, I'm talking about my *leisure* reading.

My friends never found me without some manner of text or a proportionally interesting science fiction tale.

I was blessed to have known both some actual rocket scientists[26] who had worked on the Moon

---

[26] Thanks, Ernie! A small digression here – Ernie not only showed me how math worked, (along with his son Paul, a true math genius), he also introduced me to computers. He had a Trash-80, Atari and -gasp and swoon- an *AMIGA*! Paul and I were both rapt with attention when he would fire up the *AMIGA*, and we could see graphics! Of course, he also programmed a bunch of now-lost applications. He also gave me a good idea of what happens to old rocket scientists after they 'retire'. He is sorely missed. Ad Astra...

Rocket Surgeon

Shot, and some science fiction authors and other celebrities, by virtue of the "*Star Trek* Fans United" or STFU.[27]

The President[28] of that stalwart fan club introduced me to some of the actors from "*Star Trek*" over the years.

I eventually met George Takei, Leonard Nimoy, Mark Lenard, and Jimmy Doohan at various events and after-parties.

I admired them for their depictions of fictional heroes.

And they provided me with some measure of drive – I wanted to be who they were *on TV*.

I wanted an "*Enterprise*".

So, like many of my and the next generation[29], I set out to develop the necessary skills to build one.

---

[27] **STFU** did NOT mean then what it means today...

[28] Mr. John Ellis, a very good friend of mine to this day

[29] See what I did there?

A.E. Williams

Science Degree – First Iteration

Once I had nailed high school, I went off to college, where I decided I wanted to become an Ocean Engineer.[30]

(One of the dreams I had, aside from Rocket Surgeon, was to live in an underwater city.)

It seemed like a cool idea.

Imagine, living in a fishbowl, while outside were real, live fish![31]

I thought I had the necessary chops, and dove into my local State college program to achieve eventual mastery over the oceans.

Then, a little thing called *life* got in the way.

That and my course grades...

To put it mildly, my first attempt at being a real engineer failed miserably.

I hated calculus.

---

[30] Chalk it up to too many Jacques Cousteau films, the National Geographic, and my proximity to the Atlantic Ocean.

[31] It was the exact opposite of owning an aquarium.

Rocket Surgeon

The physics teacher I had was inscrutable, and the math teacher spoke with such disdain and an unintelligible accent that I gave up, temporarily.

About that time, I lucked into another scientist, who gave me my first big break – he hired me to work at a defense contractor, in the lab.

This company made jet and rocket engines for the military and civilian markets.

Their facilities were way out in the swamps, requiring a seventy-four mile trip each day.

And, that was on those days when I wasn't driving another fifty miles to attend classes.

I spent a decade mixing work and school.

My daily routine had me rubbing elbows and sitting in rooms with some of the sharpest minds in aerospace at that time. [32]

We worked on complex problems regarding superalloy composition, corrosion, electroplating and

---

[32] This was before the *Challenger*, and we would discuss supersonic and hypersonic aircraft, satellite communications, and the performance parameters of the latest top secret jets. It was heady stuff...

coating of materials, and the thermal properties of same, while exotic fuels exploded inside closed chambers to provide energy for flight.

We tested these parts in salt-spray cabinets, ovens, immersed them in acids and other dangerous liquids, and played with electricity.

We even blew stuff up, once in a while.

Sometimes, *on purpose*!

I was informed that the *real* fun was to be had once I had graduated with my degree in engineering.

Therefore, I set my sights on pursuing that goal, and so returned to the halls of academia in my spare time (i.e. non-work).

I bent to the task of conquering calculus and physics once more.

I still thought O.E. was a good path.

However, my youthful exuberance was brought to an unfortunate end by a series of interesting events.

The first of these had to do with my work on coatings for various jet turbine parts to prevent corrosion from salt water.[33]

We would put these test pieces in a humid, hot salty cabinet for up to six months, photographing the corrosive effects of interacting with a benign substance that cover's over seventy percent of the planet's surface.

It was informative to see hard metals turn to pitted, useless junk in a matter of days, in some cases, because of a bit of brine.

This was important to the Navies of the world, as they had long ago fought the battle of corrosion with ships of the line.

Commercial vessels also had found ways of protecting their hulls, using sacrificial paints and zinc anodes.[34]

The application of this kind of science to the jet engines of military aircraft was a novel idea, and posed many challenges.

---

[33] Oddly enough, there is a LOT of **salt** water on our planet. This plays all kinds of merry hell with metals.

[34] https://en.wikipedia.org/wiki/Galvanic_anode

It was one thing to put a hunk of metal on a ship's hull, and quite another to put it inside the hot mess of a combustion chamber for an F-14 *Tomcat*.

For one thing, Foreign Object Damage, or *FOD*, would ensue.

Any material ingested into a jet engine is in for a nasty time.

Not only that, but as the now-mangled mess traverses through the engine, the engine parts usually break.

This causes more consternation, eventually culminating in a flameout, whereby the engine basically shits itself.

All the parts blow out the back, and sit smoking and sputtering on the deck behind the aircraft.

It's quite exciting to see!

It's also very expensive.

An F-14 had two engines, and these babies weren't cheap![35]

---

[35] *"Top Gun"* spent a lot of time giving the public a pretty good idea of what the F-14's capabilities were at the time. I remember seeing that plane at several airshows before I was

It dawned on me, during this time, that undersea cities may have some inherent problems that I had not heretofore considered.

Perhaps we would need to develop a better understanding of how materials reacted with other elements and how that would affect their intrinsic strengths a bit more.

Sounds like a lot of math...

Meanwhile, I had picked up a part-time gig to pay my mounting educational bills.

It was at a *K-Mart* styled department store that I met an Assistant Manager, who had obtained the exact degree I sought.

I figured maybe he was moonlighting like I was as the going rate for O.E.'s wasn't exactly a kings' ransom, but I was mistaken.

I asked him why he was engaged in sticking price

---

employed working on the TF-30. It put on quite a performance.

tags[36] onto products in the store, instead of mucking about underwater in a pressure suit.

He said it paid better than his "useless" O.E. degree.[37]

He kindly informed me that, at that time, the only places to garner work in the field were in Antarctica or the Navy.

Neither appealed to this gentleman, who liked the sun and fun of South Florida.

This proved to be information that caused me to recalculate the current trajectory of my career of choice:

1) I worked for an AEROSPACE company, not the Navy.
2) Although I enjoyed cold weather, I had recently seen John Carpenter's *"The Thing"*

---

[36] Back in the days of pre-Internet history, a small piece of paper that had adhesive backing was used to announce the 'price' or 'cost' of an item in a department store. These 'stores' were buildings whose sole purpose was to showcase 'products' that people could purchase using 'money'. Today, of course, all of that has been replaced by online shopping and Google Wallet. Or Apple... (if you are a smug purist)

[37] My apologies to anyone who has an O.E. degree that is NOT useless.

and therefore was a bit reluctant to move to Antarctica.
3) I worked in a chemistry lab (oh, and part-time stocking shelves)
4) Perhaps a mechanical engineering degree was not -too- different than an O.E., right?
5) I mean, there's plenty of mechanical engineers...
6) ...
7) Profit!

I decided to grudgingly follow my brain, instead of my heart.

But wait!

Now, I would be closer to achieving my original goal of getting to space!

Fuck the oceans!

Nuke the whales, baby!

Outer space, here I come!

Science Degree – Second Iteration

Changing to the Mechanical Engineering program was easily accomplished, but it also changed the parameters of my necessary course-work.

Now, I had to master even MORE math and physics!

As you may recall, my first experience with this did not give me much satisfaction, nor instill much confidence.

I enlisted the aid of a few of the engineers at work to tutor me, and I eventually gained a modicum of rudimentary understanding as to the mysterious ways of higher math.

In time, I conquered Calc II, and felt motivated to continue the path towards becoming a true mechanical engineering maven.

It fell upon me to begin taking some serious classes.

- Statics
- Dynamics
- Strengths of Materials
- Physics II
- Calculus III
- Differential Equations

It soon dawned upon me that I was, once again,

NOT getting the necessary grades upon which success is founded.

Oh, I did okay.

I passed most of them.

But, one day it hit me.

I was taking too long to achieve my goal of getting an Engineering Degree.

Like, hey, I had been in college for SEVEN years!

To my credit, I had almost two-hundred ... credits.

They wove all over the place.

Organic Chemistry, HVAC, Solar Radiation systems, Electronics, Programming in Fortran and C, Statistics... you name it, I probably took it.

And, I kept taking more courses until one day I woke up at thirty years old, degree-less.

I had to do something to capture all this expertise and life experience.

Naturally, I changed majors and went into a Bachelor's Degree program in Human Resources at a totally different college.

A.E. Williams

HR Degree – First Iteration / Science Degree – Third Iteration

I spent the next year obtaining a Bachelor's degree in Human Resource Management.[38]

---

[38] A momentary digression regarding financing one's education. I was very lucky (and GRATEFUL!) that the company for which I worked offered a program that allowed me to take a course at my expense, pass it, and then reimburse me for the tuition. While it wasn't a lot, it nevertheless provided me with incentive to continue my studies until I got my degrees. I was able to get an A.A., two Bachelor's and a goodly part of a Master's degree this way. Whether the time spent was of any long-term benefit is an interesting question to me. While I definitely see how a college education provides additional options, it also is evident that these credentials guarantee NOTHING in obtaining well-paying employment past a certain age. I suppose this may sound bitter, but I am not upset at having attended college. Although I can honestly say I don't feel I used the actual techniques I learned for specific tasks in my career, the overall learning experience and ability to organize projects came in quite handy several times. Today, I caution people **not** to go into any kind of debt for any reason.

As our world is becoming more globalized, and the demographic of STEM workers more diverse, it is more important than ever to choose wisely, for the average student. Being saddled with crippling debt just to *maybe* get a minor competitive edge in the job-market is not a good path for success. I do not envy today's high schoolers, and freshmen. At the exact time when most of us are at our most uncertain age, and are just beginning to figure out the way the world functions, we are trying to make good decisions, after maybe not paying

It was an interesting time, and taught me how to write well, and stay up late.

I was writing out my degree thesis[39] on a Northgate 286 PC.

This *'personal computer'* was a far-cry from the old Trash 80 and Commodore's we played with in high school.

It had a 24-pin dot matrix printer, and I would buy a box of perforated paper at Costco almost every month, along with a new toner cartridge.

And, it was **better** than the computers I used at work! [40]

I would write my thesis, and work on CD-ROM library catalog systems while on campus for my HR degree. Then, I would go out to work and do all

---

as much attention as we should have along the way - and now we get to try to understand contracts and obligations that may haunt us for decades. To say this is challenging is a vast understatement.

[39] Yes, a *thesis* for a Bachelor's degree!

[40] Seriously, for many years, my *personal* computer was far more capable than the workstations we used at a major defense contractor!

manner of classified stuff [41] using computer systems that were so inadequate that it made me laugh.[42]

After a year of this, I graduated, with honors.

And promptly failed to gain a promotion at work...

So, I transferred over to where the real fun was – *Quality* Engineering.

We were tasked with making sure the Company's purchases of dilithium crystals – sorry, steel and other mundane metals – for building jets and rockets were within the specified parameters of the design documents.

And, hence began my quest for my next degree.

---

[41] Mostly chemical processes for manufacturing, some of which led to patents.

[42] Around this time, I also took a course in PC repair, and began to program and build my own machines. Then, I started building them for friends and family. At times, it was lucrative and frustrating, and I did this on and off for about twenty years.

Rocket Surgeon

Science Degree – Fourth Iteration

So now I had a four-year degree, a huge amount of relatively useless credits, (the majority in STEM coursework), and a new job where I was working with the raw materials that went into the Space Shuttle LOX Turbopump.

Now, THIS was exciting work!

We'd sift through all these purchase orders, invoices and manifests, performing audits to assure we got what the Company paid for.

We'd run some non-destructive tests, and then analyze these test results.

If there were any variations or concerns, we'd cut up actual parts and subject them even more destructive testing.

Then, we'd collate, analyze, and have meeting to discuss the importance or consequences of the data.

It all then was reported NASA, in the hope of avoiding another *Challenger* disaster.[43]

I was actually doing the kind of work that I'd imagined the dudes and dudettes[44] from "Star Trek" were performing in their Federated, sterile and super-science-fiction worlds.

I mean, sure, they used computers that actually seemed to be useful, and not glorified type-writers.

And, of course they didn't really use paper in the future.[45]

But it was heady stuff!

---

[43] Did you know that, for the LOX turbopump on the SSME, a semi-tractor trailer load of documentation accompanied each pump? Neither did I until I had to write all that crap! Well, it was really a team of about fifty engineers that wrote it all up. And, maybe that's a bit of an exaggeration regarding the amount of paper. Oh, wait...no it isn't. In fact, when we were exploring putting it all on WORM media, it became quickly evident that we had over 17 years of data for one program. It took up an entire room of filing cabinets. We managed to condense it down to only four CD-ROMS. Of course, today all of that would fit in your smartphone, right next to your Instagram photos and YouTube cat videos.

[44] Hey, it's the Nineties!

[45] Heck, even the toilet paper was replaced! Three sea shells!

Rocket Surgeon

It was a time of deadlines, multi-million dollar delays, and beating the Red Menace for the Gipper!

Hell, it would only be a matter of time until we'd made anti-matter!

Then, I ran into the real-world wall of physics, again.

Now, please understand that this was before *"The Big Bang Theory"* made quantum physics a Twitter hashtag.

OK, so we no longer used slide rules, [46] and we had some pretty good data regarding how rockets performed.

But, there was this pesky thing called "gravity", and the idea of a space-time fabric, and oh, yeah, RELATIVITY.

I became a voracious reader of science-fiction, again, looking for clues as to how we could create black holes in our toilets, or anti-matter in our tea cups[47] but there were none to be found.

---

[46] Only to check the work of minions!

[47] Brownian motion generators on the cheap!

So, I did what any normal scientist cum engineer would do - I started watching television again, in between the last six classes I needed to get my coveted *hard* science degree.

And, that's where I found *"Star Trek"* was -still- inspiring youngsters everywhere.

Except, now the Federation had Klingons!

And something called Number One, and androids?

What the fuck?

How did I miss *this* when it was first on?

After catching up with Picard and the gang from *"The Next Generation"*[48], I found I was finished with my science degree.

And, promptly, I was transferred out of the area where I had spent four long years toiling on our most beloved national firecracker[49] and into the venue of hard-math and physics.

---

[48] By this point I was watching it in syndication, having completely ignored it on Prime time.

[49] Sorry.

This is where I learned **logic**.

Hard, unfailing, unforgiveable, intractable, merciless **logic**.

The kind that bankers use to figure out how to crush nations and send the population of the world into indentured wage slavery.

LOGIC [50].

Cold, implacable logic, the kind that the computers in "*2001*" and "*Saturn V*" used to kill off the weak meatbag humans.[51]

I was given the job to program inspection robots.

Now, this may sound cool[52] to the uninitiated, but the facts were that the job was routine, most of the time.

It led to one of my most memorable victories in life, where I saved the company a fortune in calculation time by simply importing all the data into an

---

[50] Not the Mr. Spock kind. Spock was half-human.

[51] Wow, that's a Kubrick, Kirk Douglas and '*Futurama*' reference all in one!

[52] And, believe me, it was - occasionally.

Excel™ spreadsheet and using the "Radar" graph function.[53]

This meant working with real SSME parts, going up to Cape Canaveral, Kennedy Space Center, and out to Michoud and then Huntsville to help out with the ISS program.

I was interfacing with scientists and engineers regarding all the various elements of rocket production, satellite and missile guidance, and the vast and growing computer phenomenon of the Internet.

And, each as exciting day came and went, it underscored just how far away from the *"U.S.S Enterprise – NCC-1701"* we mere humans actually were.

It was then that my enthusiasm began to fray a bit around the edges.

I saw how difficult space travel was on robotic explorers, let alone human ones.

The real dangers of just plain old interplanetary travel were laid bare in the failures we were racking up trying to glean information from one planet

---

[53] The notoriety I gained was worth an extra helping of gruel at dinner time.

to the next.

The STS was beat up every trip, with the ceramic tiles cracking and even detaching.

Not to mention, we were letting a major portion of the craft just burn up and smash into the oceans![54]

Hell, we hadn't even gone back to the Moon since **Apollo 17**!

The marvels of modern science were insignificant compared to the illusions created in the science fiction tales of our best authors.

And this got me to doing some serious thinking about science fiction vs. science fact.

Some of the questions went like this:

1) Why do we want to go into space?

---

[54] The big, orange main fuel tank typically landed somewhere offshore. It was a huge container, good for one use. (On the *Challenger* this was the part that was compromised when the solid rocket motor O-rings let the hot gases bleed out between the propellant segments). The solid rocket boosters were another matter. These came down on parachutes, splashed into the ocean, and were then retrieved and towed back to be 're-furbished'. While still costing a fortune to do this, they did provide some measure of re-use. Not at all like the SpaceX *Falcon* booster, though.

2) Why should we go there?
3) If we choose to go, who is best qualified to choose those who will go? Why? What about personal choice? Do we really need astronauts to be 'supermen'? Should we send 'ordinary' people? How can we vet and choose who is to go?
4) What happens to the planets we have visited / will visit? Are we contaminating these 'pristine' worlds? Why should we care? The Universe is, after all, big. Really big...[55]
5) Is there an economic advantage to doing space? How can we transfer that wealth to our own nations, citizens and use it for best purposes?
6) Are corporations the only way for private rocketry efforts?
7) Can a government / corporate subsidy be effective? Will regulatory constraints stifle innovation?
8) What about love? (i.e. sex in space)
9) Is Einstein right about FTL travel? Can we find a way to either side-step the space-time fabric or warp it?[56]

---

[55] Insert obligatory "Hitchiker's Guide to the Galaxy" quote. And thank you, Douglas Adams. RIP.

[56] Yeah, I know. Bad pun...sorry.

Rocket Surgeon

10) What about Artificial Intelligences?
11) What about hibernation?
12) What about Relativistic effects? Who would want to go on a trip where everything and everyone you know now will be long gone when you return? Would you ever need or want to return?
13) How can we power such ships? Are antimatter or synthetic black holes something we can hope to achieve?
14) What is gravity, really?
15) Why do similar patterns recur in our Universe, regardless of scale? How does this give us clues as to the nature of things.[57]
16) Can we download our consciousness into machines?
17) Has it already been done??? (!!!)
18) Would an A.I. go mad?[58]

---

[57] Gaseous clouds, whether under the ocean spewing from hydrothermal vents, at sea level from volcanic action, or spread across light years in nebula all appear to have the same 'shape'. Yet, they exist under tremendously different temperatures, pressures and scale. What is it about energy levels, atomic and quantum states that produce these chaotic yet recognizable patterns? Fractal math seems to hold some of the answers, but there is also an interesting facet that is gravitic in nature.

[58] Roko's Basilisk!

19) Is Time Travel something that will need to be considered?
20) What about the Faster Rocket Problem?[59]
21) Is FTL possible?
22) What about Generation Ships?
23) What about instantaneous transportation across vast distances?
24) Can space be 'folded'? What kinds of energies are needed?
25) What's for dinner? What's A.E. been drinking?[60]

I pondered these questions for a decade or so, at which point my career trajectory took a different path, into the realm of Information Security.

And, about a dozen years later, *"Terminal Reset"* was birthed.[61]

---

[59] This is where a rocket is launched on a long-term (think centuries) mission. During that time, better technology is developed that eclipses the original design by orders of magnitude. Such tech would pass by the older ship. There is a huge drag on technology development posed by this problem. Many will use it as an excuse to delay or not even build a given design, hoping or thinking that in a few decades, the new tech will miraculously appear.

[60] Maker's Mark and Classic Dr. Pepper, or Classic Pepsi, if you really care. Yay! Sugar! Boo! HFCS!

[61] With the help of my good friend, William E. Bartels

Rocket Surgeon

A lot of the questions in that list got their answers, in the meantime.[62]

But, remember that theories are *just* ideas.

Proving them requires repetition and repeatable results.

And, sometimes, they are superseded by better models of reality.

The flat earth gave way to our current understanding of things.

That led to the thinking that spawned the idea of rocketry.

That led to our landing on the Moon, and exploring the Solar System.

Quantum physics is in its infancy, to be sure, and has already given us many miracles.

Cell phones, LED flat televisions, and advanced storage for computers are all items that use our rudimentary understanding of these phenomena to

---

[62] I would point the reader to *"A Brief History of Time"* by Hawking, and maybe even *Wikipedia* for some insight as to what science has uncovered regarding those answers.

provide us with tools.

Perhaps in another century, we will finally get to those flying cars, and jetpacks.

If we are lucky, the problems of undersea living may have been solved.

Maybe some of us will be immortal, or at least have their brains and consciousness encased in cybernetic machines.

Maybe we will have finally figured out just how gravity works, and perhaps we may even have created the huge space liners of science fiction, bending the fabric of space to our will.

Or, maybe all that remains of the Human Race will be a burnt cinder, aimlessly drifting through space; the only epitaph we leave will be some metallic junk on the Moon and flying through deep space, with the human race not even missed by the Cosmos.

❋ ❋ ❋ ❋ ❋

Well, now that you've enjoyed that autobiographical rant about just why I felt compelled to write "Terminal Reset" and how that came about, I am presenting the original articles I wrote for the Speculative Fiction Showcase.

I want to thank a few people that have helped make this a reality.

First, the *Speculative Fiction Showcase* founders and editors, Cora Buhlert, Jessica Rydill, for their kind support and feedback on my admittedly personal perspective on manned spaceflight.

They've done tremendous things in helping me get my thoughts organized in a manner that is approachable to the layperson. I can't thank them enough![63]

Next, Steve Polyanchek, who created all the wonderful cover art, and the awesome video for our YouTube™ channel.

I have met a significant number of other authors over the past few years that have made things

---

[63] I still owe them a few more articles, so there may eventually be a Volume Two of this book!

much easier for me in many ways.

There are too many to list them all, but I'd like to thank Olivia, Alex, Rosemary, Scarlet, Jay, Hugh, Wayne, Ryk and, of course, Sean.

Also, much thanks to the members of the Boca Raton Meetup Group "The Inkwell" for their assistance and critiques.

John Ellis, Massad Ayoob, Rachel Galvin, Mark Vidal, Tony Wharton, Jason Foster and Michael Haney have also given me a lot of support and critical feedback, and were there through some very tough times for me.

Thank you all!

I would be very remiss if I did not also include William E. Bartels, the co-author of "Terminal Reset" for his ideas and encouragement. It is rare to find such a good friend, who has both the kind of character and the intrinsic values that exemplify the best our country can offer.

Finally, I'd like to thank my family for putting up with a man who may be pretty much crazy, but whose heart is at least mostly in the right place![64]

---

[64] For a Martian...

Rocket Surgeon

A.E. Williams

A.E. Williams

Speculative Fiction Showcase Articles

These articles were originally published in the "[Speculative Fiction Showcase](#)" blog, and are reprinted by permission.

Portions of this particular essay are taken from previous forum posts, or are excerpted from Author's notes from "Terminal Reset – The Coming of The Wave – Episode Eight".

Why MANNED Space Exploration? A Treatise on the Necessity for Manned Spaceflight
By A. E. Williams

NOVEMBER, 2014

The inherent danger of the privatization of manned space flight is in the news, again, with the recent destruction of the Virgin Galactic SpaceShip Two class test vehicle, the *"VSS Enterprise"*.

The test pilots on this flight were Peter Siebold and Michael Alsbury, and astronaut Alsbury died during the crash.

The bravery of these men has allowed humanity to boldly go, again, into the exploration of the unknown.

Three days earlier, a non-manned private mission by Orbital was destroyed six seconds after launch when it went off course.

Less than two weeks after the Virgin incident, the robotic lander *Philae*, bounced around on the surface of Comet Churyumov–Gerasimenko (67P).

Space exploration is a funny thing for humans to

undertake.

As a species, humanity has an innate curiosity about everything we experience, and a need to dissect and understand the intricacies of our world and the larger Universe surrounding it.

It applies to all aspects of life.

We want to know more, always more, about ourselves, other people, cultures, and the ways in which we interact.

We want to know about the WHY, and the HOW.

The human race is cursed by infinite curiosity, ever reaching into unknown areas.

This aspect of humanity has caused untold suffering, in that we have inflicted our wills and viewpoints, often violently, in the name of one cause or other.

People have been vilified, categorized as worthy of extermination, conquered, and enslaved over the perception (right or wrong) that they were the cause of some misfortune or other.

The expansionist and egoistic leaders of the past have forced us to evolve, to develop civilized societies that provided the foundations for the great achievements to which we bear witness today.

There is a particular selfishness amongst these leaders to be remembered by history, to be thought of as to having added to the posterity of their lineage.

Our major technological achievements are the culmination of perhaps thousands of critical minds, yet few are honored by the text books of our schools.

So, when manned space flight comes up as a subject of discussion, it stands alone in the context of being a truly human experience, shared by visionaries and workers alike.

And, in my view, that is an odd arrangement of agreement on a planet so fraught with disagreements over petty and trivial things.

At this time, the United States Space Program counts 18 astronauts as having given their lives in the exploration of space.

Of course, there are far more sacrifices than that, if one includes the scientists, test engineers, technicians and other people who were killed from the inception of rocketry, through the Cold War Space

Race, and leading up through today's corporate adoption of manned space flight.

I personally can name at least fifteen people with whom I worked who have died that were instrumental in making significant advancements in the industry.

Some died of heart attacks, some of 'natural causes'.

Some died because of the technology.

Now, I do not want to assume that the value of one person's life is worth more or less than any other.

I would like you to consider that almost all of these people, I feel, would argue that being a participant in our Space Program was worth all of the cost and effort expended, including the loss of lives.

Let me be clear – people die in warfare, millions of them.

Their lives are certainly every bit as valuable as yours or mine, or your children, or any other human.

But, for the brief time we inhabit it, most of us are bound to this planet.

Rocket Surgeon

The vast majority of humanity never goes any higher than their local hilltop.

Many climb mountains, or fly in an airplane.

But an astoundingly paltry few of us have made it into space, the Great Unknown that weighs above all of our heads for the entirety of our lives.

And that's a damn shame.

And it needs to be corrected, right now.

❋ ❋ ❋ ❋ ❋

AN ANNOYING AUTOBIOGRAPHICAL DIGRESSION

In October of 2012, I visited the Air and Space Museum in San Diego, California.

Outside of the museum sits an *SR-71 Blackbird*.

The *Blackbird* had two J-58 engines, manufactured by Pratt & Whitney, and I was privileged to have been part of the team that provided experimental data on test stand A-1 back in the early and mid-1980's.

Science Fiction Author and Jet Engine Guy

Here I am posing with the SR-71, which is a magnificent aircraft, well ahead of its time, whose performance is still unmatched to this day.

A lot of that was because of the unique airframe, and a lot of it had to do with these magnificent beasts of engines.

The SR-71 was a satellite that orbited the Earth within its own atmosphere, providing tactical data and information on the enemy forces that threatened the United States during its tenure.

It could be deployed and on-station within only a scant few hours, and, because it was flown by pilots, also had a built-in ability to deal with escalating strategic situations instantly.

The men who flew it were representative of the best of humanity - sharp, keen minds, physically excellent and dedicated to the cause of advancing our knowledge.

SR-71 Main Engine, the J-58, in Full Afterburner, Non-Assisted

This picture was snapped on the A-1 Stand[65], in the midst of the Florida swamps, from that time.

You can see these diamond shaped "things" in the exhaust.

Those are called "shock-diamonds", and occur at high speeds, above supersonic.

The engine is in afterburner mode here, which is NOT anything compared to when it ran in RAMJET mode.

The J-58 used a hybrid system that allowed normal

---

[65] Source: Kurt Schmidt Writes:
"The picture actually shows test cell A-1 at Pratt and Whitney's West Palm Beach facility. I had the pleasure of working this test cell in the mid 1980's. This test cell was actually an altitude simulation cell used for testing purposes. (We had a additional cell used for sea level runs for motors which were overhauled onsite also.) It used a non-afterburning J-79 as a slave motor. The exhaust of the slave was introduced to the inlet of the J-58 through a series of valves thereby simulating the speed, temperature, and density of the air at the inlet normally seen during flight. In this particular picture the motor is running at sea level as indicated by the inlet screen. I spent many hot and humid nights servicing, mounting, and running this particular cell. I was one of a crew of five. Some of the greatest co-workers I have ever had the privilege to of worked with. Sadly, the cell is no more. It didn't go without a fight though. The contractor had to repeatedly repair his demolition equipment saying it was the hardest concrete he had ever had to remove. Sincerely, Kurt Schmidt"
http://www.wvi.com/~sr71webmaster/j-58~1.htm

aspirated flight during take-off and landing and climb to altitude.

It had a mechanism that would move an aerospike inside of the engine, that could manage where the supersonic shock waves would occur, and, at altitude, by-passed the turbine section to operate fully as a ramjet.

BTW, I am in this photo.

Now, if you can't really see **me** in this photo, it's because I am in the blockhouse, *behind* that engine - with my hand on the throttle.

Let me tell you, it was something else.

That raw power, unleashed in front of your eyes, watching those numbers on the Mach gauge climbing to **CLASSIFIED** levels!

The ground rumbled, the building shook, and it felt as though an earthquake were happening right there!

Only, if I reduced power, it didn't shake quite as much, and if I gave it MOAR POWER it really started to loosen your fillings.

Now, I am not saying this to brag about the experience.

I want to share the excitement and feeling of awe that I felt to be there, at that moment, which was the culmination of hundreds of people's efforts, testing, calculations and sweat.

Some additional annoying autobiographic info:

The F-14 *Tomcat* used the TF-30 afterburning turbojet engine, and was one of the premier aircraft ever created, with a movable swept-wing for optimal performance across its flight envelope. It was the 'real' star of the movie "**Top Gun**".

The F-15 *Eagle* and the F-16 *Falcon* are two other aircraft in the inventory, that perform different functions. The F-15 is an air-superiority fighter. The F-16 is for close air support (CAS) and also is a very capable fighter.

The F-15 has the distinction of being the only aircraft to shoot down a satellite[66].

---

[66] https://en.wikipedia.org/wiki/Solwind

The F-15 can stand on its tail and accelerate vertically in a climb to over 25,000 feet.

The F-15 and F-16 both use the F-100 engine.

The F-15 has two of them, the F-16 only one.

The F-16 is, affectionately, known as the **'Lawn Dart'**.

Think about that for a moment...

The F-22 Raptor is probably the most advanced aircraft flying today, (as far as the public is aware, at least).

The F-22 uses two F-119 engines, which are derivatives of the F-100, and much more powerful.

As well, they have interesting vectored nozzles.

I was involved in setting up production areas for these.

All of the engines powering these planes were built by P&W.

And, I worked on all of them, in one capacity or another.

To bring us full circle into the realm of manned space flight, I would like to introduce the STS, or Space Transportation System.

There are three Space Shuttle Main Engines used on each STS flight.[67]

Each engine has a LOX and Fuel turbopump for that engine.

All of this technology was developed from decades of hard work.

People trained for years to develop the skills needed from math, science, engineering, chemistry, and physics disciplines.

There were hundreds of people involved in design, testing, materials science, aerospace engineering and physics.

Thermodynamics, statics, and physical chemistry all lent a hand in determining how best to create these marvelous machines.

There were architects, construction workers, plant

---

[67] The turbopumps did NOT fail during the Challenger mission. That was a solid rocket booster O-ring failure.

safety and industrial security, electronics and electricians, custodians, clerks, accountants, contract specialists, government liaisons, and more than I can list, all involved in just building the facilities where these fantastic creations could be built and tested.

There were shipping and receiving areas, warehouses with forklifts, trucks, aircraft, computers, mainframes, and test areas.

At one point, over 8,000 people were involved in manufacturing, testing and getting the engines to where they were needed, just at one plant!

There were plenty of mistakes, and outright failures, such as one of the turbine disks exploding during a test and puncturing a containment vessel, then shooting off into a wall in a bathroom – over a hundred feet and three rooms away.

I put in a pretty decent amount of time, working in many disparate areas. There was a huge amount of secrecy, security and record keeping.

So, when I start off telling you how interesting, exciting, dangerous and absolutely mind-blowing it is to have worked on these programs, I hope you can understand my passion.

A.E. Williams

I hope to make it very clear why I feel so strongly that manned space flight is maybe the most important thing EVER that humans have undertaken.

And, I also hope you can understand my excitement that manned space flight is entering a new phase with the opportunity of having private corporations engaged in continuing the tradition set forth in the last five decades by governments.

That is trivial adventure compared to what is a routine daily mission to our best.

Many years ago, in the original *"Star Trek"* series episode "*Return to Tomorrow*", Captain James T. Kirk outlines to the crew of a mythical starship, (also christened **ENTERPRISE**), the importance of accepting that risk.

You can view this inspiring speech here.

It was written by John T. Dugan, under the pen-name "John Kingsbridge", and is probably the best and most succinct reason anyone involved in space

Rocket Surgeon

exploration can give as to "Why?"

That speech has always stayed with me, and is one of the reasons I ended up where I am.

The facts are the same now as they were then -- when you have complicated machines, thousands of people involved in constructing and launching the most advanced mechanisms mankind has ever created, and the inevitable constraints of cost and time, you are going to have to assume some measure of risk.

Humans have accepted that risk over the centuries, which is why we stand at a pinnacle of great achievements today.

Our current world is a fast-paced, interconnected global powerhouse of talent and intellect.

Every day, advances are made towards creations that free the majority of us from worries about survival, including clean water, sanitation facilities, and food.

Yes, we still have a very far way to travel before we have alleviated these problems for all of us, but I would illustrate that those problems are ones of political and not technical nature.

We have the ability to feed, clothe, house, educate and care for everyone on the planet.

We do not seem to have the necessary maturity to rise above prejudice and bias about each other – yet.

A manned space program provides that vision, and the necessary maturity to do the hard things.

There is no margin for error.

The attention to detail is ferocious, and the consequences permanent.

We learn, we try, we fail, and we integrate the lessons into our future endeavors.

People and equipment are lost.

Decades of work disintegrate as we all move into the future, and the hard-won secrets are sometimes lost as well.

Above all, we have had the privilege of watching the best of us test the bounds of Earth, touch the Heavens, and leave footprints on other worlds.

It is for this reason alone, the celebration of the hubris of humanity, that we need to continue to reach out, accept the risks and continue on our journey of manned space exploration.

"Ad Astra Per Aspara."

MANNED Space Exploration? It's Getting Old – PART 1

By A. E. Williams

JANUARY, 2015

This is the year 2015.

Based on the initial excitement to which I was exposed as a young lad during the Sixties, I was expecting to see flying cars, jet packs, robot maids and round trips to our space stations, Lunar bases or even Martian cities on a daily basis.

The ideas tossed about haphazardly by Hollywood and the fledgling television industry sought to assure all of us that Science was entering its Golden Age, and what were mere figments living in the imaginations of the much-vaunted science fiction writers of years past were coming to fruition.

We had lasers, and satellites, and telescopes that were probing the Universe. We had the atomic bomb, and also the nuclear submarine.

How long until the nuclear bomber or nuclear oil tanker?

Rocket Surgeon

Manned space exploration was the gatekeeper of these wonders.

Research into keeping humans alive in space caused simultaneous expansion in the fields of engineering, medicine, communications and rocketry.

Von Braun and the team gave us the engines of our Future, the means to conquer the Solar System and then the Galaxy.

It was Manifest Destiny writ large, with the entire Human Race the heirs to a great hegemony of...of...

Well, horse feathers.

We didn't manage to get even close to that dream, now did we?

I mean, we don't have lunar bases, underwater cities, space travel to and from our space stations, or even a decent atomic rocket.

There seems to be a dearth of jetpacks in my neighborhood.

I do see the flying car has been making inroads in some countries, but the myriad problems of airborne traffic jams are still years away from being solved.

Google has been working on getting us to become passengers, rather than drivers of our land-based (and presumably sea-based) vehicles.

One can imagine that there will be an extension of some manner to incorporate this into a personal aerial vehicle at some point.

In spite of the fact that we have not received all of these promised wonderful gifts as a result of technology and science, we at least now have the ability to talk to each other on a global basis through small devices that we hold in our hands or put in our pockets.

Food production is an all-time high, and medicine has provided many breakthroughs that reduce the amount of potentially lethal viruses and infections.

As we grow older we notice the things fall apart and tend to decline.

This is true regardless of whether it is a building such as the pyramids, us, our pets or our children.

As a science fiction writer, I tend to notice a lot of degradation in systems, or what Rudolf Clausius called *entropy*.

Entropy is the tendency for the universe to reach a

steady-state, one where the energy of the universe is evenly distributed.

In physics, Sir Isaac Newton invented calculus - the mathematics needed to understand and analyze the mechanics of this situation. Newton proposed what is known as the **Second Law Of Thermodynamics**.

When entropy reaches its maximum the system is in a steady-state and in maximum disorder.

Chaos.[68]

This loss of energy in a system, and the increase of entropy, is the exact opposite of what everybody wishes were happening, which is increasing order and stability.

In theoretical physics this tendency to stave off chaos is desired, and so energy is added to these systems to delay the inevitable.

Real world issues tend to illustrate that, despite

---

[68] Professors Stephen Hawking and Jacob Bekenstein have done work to show that black holes are the areas of maximum entropy in our universe. Currently, though, even Hawking is undergoing a rethinking regarding this, because of new discoveries in understanding quantum mechanics. Source.

our best efforts, Nature has a way of not complying with our wishes.

From the social implications of terrorist attacks and government monitoring and surveillance of their populations, to the very real decay of our infrastructure, in many ways entropy is indeed increasing to the maximum.

Another contentious area is how our existence and the technologies we are developing are impacting the ecosystem of our planet.

In spite of our best attempts to thwart it, entropy marches towards its ultimate conclusion – The Heat Death of the Universe.

Science fiction authors tend to project, exaggerate and lampoon these subjects for good purpose in their writing.

There is some vicarious and arcane satisfaction and entertainment to be had dancing on the grave of the world.

Whether that end is by asteroid, comet, volcanoes, or some other weird phenomena thrust upon us from Outer Space™, we all stare in rapt fascination, like a deer caught in the headlights of an approaching vehicle.

Rocket Surgeon

This decrepitude, this march toward Oblivion, is part of what drives us, the Human Race, as a species to propagate and reproduce.

Some philosophers have opined that this is the primal driving force for all of us, and that there is a symbolic penetration of the Void that we seek to experience with the most phallic of objects, the rocket ship.

But, as science-fiction authors observe the world around us, incorporating truth and myth into our stories, it seems we are only the Greek Chorus singing to the audience, as perceived by many people.

Thus, it is indeed unfortunate, but also inevitable, that these systems that are very complex are going to suffer from the passage of time.

Let's look at an example, shall we?

After World War II, the German rocket scientist Werner von Braun was brought to the United States to spearhead the science projects that would take United States to the moon.

As head of what eventually became NASA, he took the knowledge that he had developed for the Nazis at *Peenemünde* and then modified it for peaceful purposes.

The same sword that was responsible for the deaths of thousands during the World War II V-2 bombing campaigns was beaten into the plowshare that became the foundation of the United States space program.

During this post-war era, the Russians were also developing rocket technology, aided-in-turn by their captured Nazi scientists.

This led to the notorious Space Race, whereby it was decided that whoever could militarily control space would control the strategic high ground.

"Sputnik", launched in October of 1957, was the inaugural shot in that battle.

It was the first time a satellite had been launched into orbit around the earth in such a manner that some use could be made of it.

It so terrified the United States government that they accelerated the development of their own space projects.

As one obstacle after another was overcome, the Russians and the Americans were able to launch cosmonauts and astronauts - actual people - into space.

The game being played was the penultimate **King of the Hill**, with the Heavens as the prize.

But, what of those programs **today**?

What has happened to all the buildings, the plans, the blueprints and documents, the machine tools, the engines, the test equipment?

Where are the *records*?

If history is to be any indicator, we are sorely unappreciative of the Giants on whose shoulders we stand.

The fact of the matter is that almost **all** of our space technology is obsoleted mere **months** after it's developed.

A case in point recently was illustrated by a **"Physics Today"** magazine article about the new engine test facility at the **John C. Stennis Space Center** in Mississippi, (which is near Slidell, Louisiana).

I had the honor of attending some presentations and investigatory meetings regarding the STS space shuttle back in the late-1980s (and again in the early 1990s) at that facility.

At that time, Stennis and the nearby Michoud facility were engaged in the production of the STS SSME rocket engines and main fuel tank for the space shuttle.

We were tasked with finding ways to improve the workflow processes for manufacturing these parts.

The best library science and document management technology we had were based on software and computers that are dinosaurs today.

The serious nature of this business was focused on the safety concerns raised by a single unwarranted disaster – the complete destruction of **STS-51-L**.

The *Challenger*.

The 'advanced' methods used to manage the complicated processes behind the most complex vehicle ever built by man - the space shuttle - involved hundreds of people running back and forth to skim through ring-binders of printed manuals that were kept in buildings located far away from the operations center.

So, for instance, if a part was showing indication of a problem during the countdown, a phone call was made to this document storage warehouse.

A research assistant or scientist would go scurrying off to sort through the myriad binders and find as much information as possible to convey to the engineering team, by making copies from the reams of technical papers.

If there were a problem detected, we took solace in knowing that the procedures were at least written down to address it.

But, the retrieval time for these documents and binders could take many hours (or days, in the worst case).

We were exploring the use of computers to share electronic versions of the documents, stored as TIFF images on either CD-ROMs or magnetic optical drives.

The problems surrounding the definition of proper indexes, storage capacity and search algorithms, although seemingly trivial, were thorny issues back in the day.

These systems were to be networked with the main facility to assure that the engineers and launch crew would have access to these documents, without having to rely on someone at the other end of a phone line, or the need to have to walk back and forth between buildings, thus causing additional delays.

Remember this was when the Internet did not exist in its current incarnation, and the tremendous instantaneous retrieval capabilities of the 'Net and advanced searches that we enjoy today did not exist.

We did not have a Google.

We did not have an ability to pull back data instantaneously, let alone analyze it as efficiently as we do today.

Yes, we had spreadsheets and calculators and slide rules, but all of these things had to be cross-checked and agreed upon before any action was taken.

Thus, the delay resulted in many millions of dollars being wasted. In addition, many man-years of time were lost because of insufficient processes to enable having our information literally at our fingertips.

Copies would not always be of the latest revisions.

Engineering data would cause alterations to procedures.

Documentation was always running behind the actual experiment or programs.

It is laughable, now, to think of tens or hundreds of trained scientists and engineers running around with carts full of paper binders, from office to office, to meeting rooms, all trying to co-ordinate their efforts.

We truly take for granted the ease with which we can now communicate.[69]

Today all of that information is gone.

A lot of it was buried in landfills, and much more of it was incinerated because of security concerns.

But, there are still some occasional trophies unearthed.

One of the F-1 engines from a Saturn - V launch was recently retrieved from the bottom of the Atlantic Ocean off the coast of Florida, and it is one of

---

[69]Personally, I find it amazing that so much of these important communications today are about the antics of felines.

the few examples we have that *work*.

It is ironic indeed, that the best of our efforts lie at the bottom of the ocean where they were carelessly discarded almost 50 years ago.

The means to re-construct or even understand the science behind them is at risk of being lost to our inertia, and our inability to understand the minds of those pioneers, even with the miraculous machines we use mindlessly every day to preserve them.

Currently at the **John C. Stennis Space Center** facility in Mississippi, the problem is that the Constellation program has been canceled as a result of ongoing fiscal delays.

NASA, being a government entity, is fraught with the bureaucratic obstacles of any organization of that size. In February 2010, President Obama was able to get enough votes to gut the Constellation program and instead NASA has repurposed the effort.

During the heyday of the Constellation program, a facility to emulate the airless vacuum of space in which to test our latest rocket engine designs was constructed at huge cost at Stennis.

The program being revised has now led to the decision to *mothball* that facility!

What does that mean?

This building and its support facilities are now a $349 million dollar boondoggle[70].

According to a recent report, it is costing in excess of $100,000 per year to just have it sit there, in storage, waiting for the next round of financial legerdemain to allow its use. If ever.

The hard facts are that obsolescence, decay and entropy are inevitable.

That facility, even now, is subject to the ravages of time.

Every day that it sits fallow, it is approaching a point where its functionality will have diminished to the degree that no amount of effort or money will suffice to bring it back to the necessary operational specifications.

---

[70] A **'boondoggle'** is a science and engineering term for a colossal waste of taxpayer money, often for the purposes of entertaining congressional representatives at lavish soirees, and for no really good reason.

The real tragedy is that this did not need to happen.

The outlay for the manned space program is a literal financial drop in the bucket of the United States budget, and one can only stare dumbstruck when contrasting the enormous expense associated with how many bombs and tons of military ordnance that have been expended during the recent overseas hostilities.

(Not to mention the entire military infrastructure, and its associated costs.)

We, the human race, MUST develop more robust techniques and systems to manage these kinds of scientific triumphs, if we are going to succeed in making routine spaceflight a reality. [71]

The risks that our globalized world faces, such as the very statistically likely impacts with comets or asteroids that create extinction level events, or the reality we are now facing from the poisoning of our home world through neglect and rampant idiocy, only serve to underscore that we are currently

---

[71] See my previous article here for why the continuation of manned space flight is not a negotiable option for our species.

stuck on a rock with no way to get off.

So, we must find ways to improve the systems.

Are we doomed, though, because our history is falling apart around us and turning into sand as we speak?

Perhaps there are other options that will guarantee our survival?

After all, science fiction is rife with the ideas that have led to actualities, as we have seen.

Interesting concepts that stem from the science fiction writer's imagination include the possibility of harnessing artificial intelligence or transplanting our intellects in ways that would enhance humans.

Such methods could make long-term space flight feasible. Such methods could allow us to seed the stars, and enable a legacy for our species.

If we are able to put aside some of our major differences about our attitudes and focus on the **real** challenge facing us, interplanetary and intergalactic migration, then we may have a chance.

We will talk about some of those issues in PART 2.

NEXT UP:

February - MANNED Space Exploration? It's Getting Old – PART 2

March - How Spaceflight and The Challenges Therein Have Been Addressed in Science Fiction.

April - An Exploration of the Physics Behind Faster Than Light Travel.

May - Cyborgs, Artificial Intelligences, Trans-Humans, the Singularity and the Merging of Humans and Machine.

June - The Physics of Science Fiction Weapons.

July - The Reality of Living in an Undersea City.

A.E. Williams, January 12, 2015

MANNED Space Exploration? It's Getting Old – PART 2

By A. E. Williams

JANUARY, 2015

In PART 1 of this article, I was pontificating about how we, as a species, are wont to let a great many of our most fabulous and wonderful achievements rot and decay, instead of trying to preserve them for posterity.

In PART 2, I would like to look at some historical documents, and also to the ways in which science fiction both predicts and extrapolates science fact.

But first, a word about record keeping.

We all know that human beings love to keep records of their achievements. We can see the monuments to great men and adventures.

The invention of writing has allowed a vast transfer of knowledge across centuries.

Science, mathematics, history, and cultural data are being saved and passed on to following generations, with a hope that lessons be learned and mistakes avoided.

We point to museums and libraries as repositories of knowledge, and the resting place of many of the great intellects of history.

But even these vaunted and revered places can suffer from the vagaries of the human condition as we swarm over the planet, seeking to master the 'Others' whom we encounter.

From Alexander the Great, to Ghengis Khan, to Napoleon, to Hitler, from Stalin to Pol Pot, to Mao, and even including recent leaders such as the Prime Ministers of England and the Presidents of the United States, men have unleashed the dogs of war.

The devastation wrought on the temples of science and engineering, to the coliseums of rational thought and reason has retarded our progress as a species far too often. [72]

We can mourn the loss of many of the best and most thorough works of mankind as we uncover evidence of purges throughout recorded time.

The Great Library at Alexandria, the Dead Sea Scrolls, lost books of the Bible, the oral traditions that never made it into a printed form and other magnificent and intricate bits of arcane knowledge are lost forever. [73]

I would not want to be perceived as being Eurocentric in compiling these examples, for there is also ample evidence that many scrolls and books from the Oriental, Persian, and Ottoman (along with

---

[72] Ironically, it is in seeking to improve ways in which we kill each other that much of the truly useful advances have been made.

[73] Not to mention the loss of artistic masterpieces and other artifacts succumbing to the senseless brutality of unfettered warfare.

many other cultures) have been lost to time.

When wars, natural disasters or pogroms decimate our pooled intelligence, it takes us a long time to recapture these experiences, and then even more time to assure that they can be passed on to our heirs and descendants.

The written word, and the invention of the printing press, the greatest advancements of their time, are inadequate to the task of easily conferring the wisdom of experience, rigorous methodology, and the testing results that can only be found by hard work and years of analytical thinking about the complex processes we need to raise ourselves into the light of rational thought.

Library science and systems management give us the tools to begin to index and develop massive taxonomies, whereby we can categorize, sort, store and retrieve this important data.

But, the march of progress brings with it the loss of

compatible media formats.

Who among you remembers the illustrious 1.44MB floppy disk as anything other than a relic?

Yet, entire businesses were founded and fun on this technology.

More recently, think of how the CD-ROM has been antiquated by the advent of cloud storage.

Soon, physical media, of the kind that an individual would own, will be a thing of the past.

All the data will reside in the massive storage systems of the Cloud, or whatever its future incarnations will become.

❊ ❊ ❊ ❊ ❊

The creation of the Internet has come as close as

anything so far invented at reaching the redundancy necessary for humanity to actually push forward our combined knowledge.

The computational algorithms, coupled with the advances in search and analytics technologies, will give future generations the ability to decode much of our present world, through cross-correctional techniques that go much further in reducing bias and prejudicial influences.

We are still in the infancy of this process, but I foresee a time in the very near future where certain standards will be adopted, and agreed upon as being sufficient to provide repeatable proofs of claims.

When that occurs, our race will hopefully have achieved the maturity to rationally deal with the responsibility that will entail. [74]

---

[74] *Wikipedia* is a start, but there are many problems to still be surmounted with providing canonical, properly cited, and

Rocket Surgeon

Let's take a closer look at how the contest between the two super powers of the 1960's shaped this process, and led to the development of a superior technology and science. In order to easily explain this situation, I have found it of value to review some of the films that were made concurrent to the Space Race.[75]

The 1967 *spy* film **"You Only Live Twice"** is a perfect example of a Cold War adventure movie at its finest, and the lengths to which Hollywood believed governments would go in order to capture

---

fully peer-reviewed information.

[75] I personally find it humorous to watch movies from the 60's that depict what Hollywood thought was accurate with regards to space travel. Many of the films show some creditable effort at maintaining realism, but others border well over into the realm of sheer fantasy.

important technical information about enemy space projects.

In that movie, the evil organization **Spectre**, (which is standing in for the USSR, the Evil Empire of that time),is led by Ernst Stavros Blofeld, (who is standing in for Nikita Kruschev, the world's Premier shoe-banger of the time).

Blofeld has ordered a special rocket be launched from a secret base located in the interior of an extinct volcano in Japan.

In the beginning montage, **Spectre's** spacecraft opens its mechanical maw and captures an orbiting Gemini space capsule. It then fired its retro rockets, and returned the unfortunate (American) space explorers back to the secret evil layer. [76]

---

[76] The science behind this bit is remarkably accurate, aside from the fact that one ship eats the other. (Parenthetically, this particular James Bond film is the source for the plot behind the enormously hilarious **"Austin Powers – International Man of Mystery"** movie).

My point here is that the technology that was depicted in the film was not too far removed from the reality of the times.

The physics was sound...if not the plot.

The film showed the dreams of Von Braun writ large – a single-stage-to-orbit craft, with the ability to take off vertically and return for a vertical landing at the initial launch site.

The recent launches of the Falcon X and Falcon 9 rockets from SpaceX demonstrate that SSTO craft are *almost* a reality.

These current technology rockets take off and land, ostensibly in the same location with the use of landing legs.

Note that this was successfully *predicted* by a James Bond film, over 40 years ago!

These film makers took pains to extrapolate a plausible scenario, to a logical conclusion.

It is science fiction of the hard variety, at its heart, but the technical difficulties regarding materials science and the actual physics of space travel have postponed the realization of this imaginary craft until today.

There are very good reasons for this.

The rockets built by Von Braun and his team at Peenemünde were designed for a one-way trip.

The fictitious rocket in the Bond film was designed not only to return to its launch base, but also to make repetitive flights with very short intervals between them, in order to replenish fuel and oxygen.

The real-world analog to this, the STS Space Shuttles, took months of careful assembly, testing and retesting to assure the astronauts and technical personnel are safe during the mission.

It does not always go well, and in 1986 and 2003 we learned again the high cost of mistakes.

Space travel is an unforgiving mistress.

The reuse of ships is hampered because of the caustic and dangerous nature of the propellants, their corrosive abilities on the materials used for containing them during conflagration, and the enormous stresses built up during the flight phase.

All of these factors provide the inevitable degradation of equipment and the breaking down of key components.

An example I can supply surrounds the turbo pumps for the SSME.

When initially designed, they were to have lasted without retrofit for 4 or 5 flights.

That is, the six turbo pumps (two for each of three main engines) would only need overhaul after four flights.

This was the design specification.

In reality, they needed to be removed and repaired

after almost every flight. [77]

If we contrast this kind of real-world performance vs fictional performance, we become aware of the myriad factual difficulties with which science fiction authors must contend.

Now, let's use the example of a *science-fiction* television series, such as **"Star Trek"**, which was made at the same approximate time.

We notice that the **USS Enterprise** is a space-warp driven craft that has a complement and crew of trained scientists, and is roughly the size of a battleship. It travels faster than light with no problems whatsoever.

---

[77] The team on which I participated managed to produce an improved LOX turbo pump, which flew on every mission after 1996 or so, until the fleet was retired. My understanding is that these particular pumps were able to weather 12 missions without need of repairs.

When the crew of the **USS Enterprise** wants to perform a translation of men or materials from ship to planet, they use the *transporter*.

This has been shown, via some interested physicists performing calculations, to be a device requiring a substantial amount of energy to operate.

Their other alternative was space flight using a shuttlecraft.

The United States Space Shuttle **Atlantis** would never be as efficient in using energy to get up and back from space as the entirely fictional shuttlecraft **Galileo 7**.

The reason behind this is the hard wall of **physics**.

To review my points, in one instance Hollywood was abiding by the laws of physics and extrapolating a plausible and likely outcome of the near-future space technology.

A single-stage-to-orbit (SSTO) vehicle was an

achievement many of the scientists and engineers of that era felt was well within the scientific abilities of the day.

As of this writing, we have yet to become adept at manipulating whatever forces of nature would allow the use of space-warp technology, transportation that is instantaneous between two points such as is depicted by the transporter, or even the regeneration machines that they use on board the *Enterprise* to make coffee.[78]

The divide between science fiction and reality is nowhere more evident than in the pristine nature of the future as depicted in the Federation and

---

[78] 3-D printing promises to bring some of that old science fiction magic closer to us all the time, though!

other films of that type versus the reality of our own space program.

The general cleanliness of the ships and cities, the way everyone has their own place and function and the way that there is unquestioned obedience to society is a bit scary.

And, let's also not forget how the technology has altered the world. People can choose to pursue old-fashioned activities such as growing grapes, or painting, or acting.

But these are at most an avocation, or hobby.

The promise of the Federation is that everything is there for a citizen of Star Fleet.

All they need do is comply.

Every time data is needed for retrieval on the show, one of the characters goes "Computer" and is almost instantly given the answer to any query, no matter how mundane or complex. [79]

If only such tremendous ability to store and retrieve vital information were true![80]

Some time ago, famed science-fiction author Larry Niven had pointed out that the original blueprints for the **Saturn – V** rockets (our **only** true spaceships of any note to date that qualify as interplanetary) were basically molding and decaying at the bottom of a filing cabinet in some office in an obscure

---

[79] The hubris of the writers to simply assume such an advanced artificial intelligence would be at the beck and call of mere humans on a star ship is pretty amazing! IMHO, the more recent film "*Interstellar*" has a very balanced examination of using A.I.s as our 'slaves'.

[80] You must understand the difference between a SEARCH engine, such as Google, and an AI. The AI is a virtual genius, with the ability to parse, understand, and correctly interpret a query regardless of from whom it is being made. There is an innate assumption that the people who ask these questions are ALLOWED to ask them. Imagine, if you will, a child asking about how to make a nuclear missile, or how to procure a phaser. The implicit understanding is that there are safeguards, right? Well, WHO makes the rules as to whom these rules apply, hmmm? **Star Trek** is pretty neat, but there are definitely some big questions that are glossed over routinely. At least you know who the bad guys are in **Star Wars.**

building in central Florida.[81]

A brief perusal of even the last 40 years' worth of technology that is being stored at Cape Canaveral in Florida shows the decrepitude and general state of disrepair of what was once the highest of all of our technologies.

Can we actually stomach the idea that this, an accomplishment that is probably the single greatest achievement of the human species, the literal zenith of human technology and achievement, is being left to rot and turn to dust?

And further, that with only a modicum of effort we could have been prevented this obscene outcome?

So, here we are facing a conundrum.

We are attempting to move into the future but the past is rapidly removing our successes.

---

[81] There may have been copies of these blueprints in Houston or California.

The victories that we had are now hollow in the sense that we have nothing to which we can point that qualifies as a suitable replacement.

One would like to imagine that something such as the privatization of space travel, coupled with the commercialization of and / or the inculcation of tourism-related space travel as being put forth by *Messieurs* Branson and Musk, among others, would lead to an appreciation of the value of such endeavors.

At what point will it become necessary to jettison this outdated information?

The question is important, in that, should mankind succeed in colonizing other worlds, it may be that the answer as to how best exploit these new resources is in our history.

The lost past may have provided the keys to our new future, and we just did not deem it important enough to preserve.

A final thought — what legacy are we leaving the future generations regarding all of this historical information?

At what point should we begin to instruct our children as to the necessary skills for survival of our species?

Do we even know what languages to use?

Mathematics, the sciences, and engineering benefit from centuries of repeatable testing and observation of our universe, and the ways in which interactions among the elements contained within happen.

Yet, as the information expands to capacity, where do we draw the lines?

Are only the smartest of us to reap the benefits of these technologies?

Do we need to know certain phrases or rituals to be allowed access to the bounties of the secret sanctums of routine knowledge?

As we mature, and reach out into space, I would expect many of these questions will be resolved.

But, we may not like the answers.

A.E. Williams

January 31, 2015

Rocket Surgeon

How Spaceflight and The Challenges Therein Have Been Addressed in Science Fiction
By A. E. Williams

MARCH, 2015

The very first science-fiction story that is popularly recognized as having anything to do with the concept of modern space travel is probably Jules Verne's "From the Earth to the Moon".

In this tale, the adventurers travel to the Earth's Moon in a modified cannon shell – it is pointed at the Moon and fired, with much fanfare, from Tampa, Florida.

Oddly enough, Verne predicted much of the issues with which manned space explorers would need to contend.

With the exception of the fatal effects of the immense shock from an instantaneous acceleration to escape / orbital velocities, Verne's idea was sound.

In fact, NASA and Iraq had programs to launch satellites using cannons as recently as 1990[82].

---

[82] Source: http://www.astronautix.com/articles/abroject.htm

Of course, these satellites would have had few moving parts; live payloads were out of the question.

But, Verne's vision sparked the idea of men traveling across space to other planets.

Since that time, there have been thousands of different tales of space adventure.

Science fiction has brought us every manner of device and apparatus to move people from one end of the galaxy (nay, the Universe!) to the other.

Some of these schemes were ridiculous, and played for satirical purposes, or were parodies of actual ideas.[83]

What I'd like to explore in this article, however, is how the very real problems of manned space travel were 'solved' to some extent by the speculative fiction authors who became very clever in just how we should proceed to move into space.

---

[83] Most of Douglas Adam's *"Hitchhiker's Guide to the Galaxy"* is about how one could travel the length and breadth of the known Universe, in time, space and hyperspace by the use of the clever 'electronic thumb'.

Rocket Surgeon

✳ ✳ ✳ ✳ ✳

Let's start with a few of the more famous modes of space travel:

Rockets –

Rockets follow the simplistic physical laws of ballistics, and sustained propulsion of a cylindrical tube filled with air, food, water and people has actually happened!

In science fiction, post-Verne, there are many places where the possible issues were enumerated and addressed.

Arthur C. Clarke's writings, including '*2001: A Space Odyssey*', Robert Heinlein's stories, E.E. 'Doc' Smith and many Golden Age authors took serious engineering minds and bent them to the task of answering thorny questions such as what actually happens to the human body in space, the effects of temperature, pressure and radiation on living entities, and the stresses of acceleration and deceleration.[84]

The mechanical engineering problems surrounding structures and forces were incorporated into many of the best hard science fiction of the times.

The authors were serious in considering what actually might occur during these flights of fancy, and came up with ingenious ideas.

Heinlein went so far as to show how the Moon could be used as a launch pad for Earth-bound missiles made of mined moon rocks, and how they could be used as serious weaponry against the Mother Planet.

His calculations were intended to show the scientific rationale about gravity wells, but he inadvertently illustrated one of the biggest issues facing spaceships, which is how to avoid debris while traveling around.

But it also showed the feasibility of using the Moon

---

[84] An interesting article could be written concerning gender roles and how they were presented by these authors. I may do one in the future! The general consensus seemed to be that women and men were equally subject to the laws of physics, though.

as a base of operations for advanced space missions.

'*Destination Moon*' was a film that used Heinlein's musings to bring some verisimilitude to the silver screen. It showed how action / reaction would work in zero gee environments, as is shown when one of the key characters uses a fire extinguisher to fly around the room.

Other '*Invasion*' films utilized stock footage of V2 and Redstone launches to convince audiences that these were actually alien craft, or were threats to other planets.

To say that the fictional creatures from other worlds were not much amused by our antics is putting it lightly.

Flying Saucers and Generation Ships–

The 1950's and '60's presented us with alien onslaughts on all sides, from the novel "*Starship Troopers*" to the flying saucers in the films *"The Thing from Another World"*, *"Mars Attacks!"* and *"Earth vs the Flying Saucers"*.

A strange variation of this was *"The Day the Earth Stood Still"* wherein Klaatu, a traveler from another galaxy (!) comes to Earth to warn us of our hubris at combining nuclear bombs and rockets.

His vehicle, a true flying saucer, was discussed in detail during some exposition in the film with the top mind of that Earth.

The relativistic effects of Einstein's new theories on Faster-Than-Light travel may have made viewers heads spin, but the dialog was grounded in scientific roots.

These stories and approaches still mainly glossed over the incredible distances involved. Even "*Forbidden Planet*", with its revolutionary saucer-ship didn't really clearly depict the time and space parameters that we now are just beginning to understand.

To get around the problem of the actual flight-times becoming lethal, the concepts of '**Generation**' ships were introduced.

These were miniature worlds, entire ecosystems with populations that traversed the blackness at a relatively slow speed, but taking millennia to get to their final destinations.

Again, Heinlein, in "*Universe*" set the bar very high.

The story took place on a giant spherical ship

where the radiation shielding had partially failed.

Mutated beings mixed it up with the normals, while the ship headed endlessly into deep space, its original purpose lost to the ravages of time.

Keir Dullea, of *'2001'* fame, explored this more fully in the television series *"The Star Lost"*.

Bruce Dern touched upon the idea of isolated ecosystems orbiting in space in *"Silent Running"*, Niven and Pournelle spoke of *Ringworlds*, and good old dependable A.C. Clarke's *"Rama"* capped off the idea until the advent of *'The Borg'* reignited it.[85]

Because it was taking so MUCH time to get from point A to point B in these stories, the science fiction authors next needed to come up with ways to

---

[85] The Generation ship is actually an idea stolen from our own human experience – we travel on such a device, every day we are alive. Think about it – a closed spacecraft that takes millions of years to travel the Universe. Food, water, life are all balanced – carefully taking billions of changes into consideration. The Earth is a star ship, but we don't really notice, since we are always imagining new and improved ways to do things.

speed things up a bit.

Enter Faster-Than-Light travel, or FTL.

Now, E.E. 'Doc' Smith had handled this quite well in the *"Skylark of Space"* and *"Grey Lensman"* space operas, by simply annihilating copper bars atomically, releasing all of their innate power into machines that manipulated bands or frequencies of this energy.

Much like radio, television and microwaves today are used for many purposes from heating food to allowing us to read words on fluorescing screens, the heroes of these adventures took all of it in stride.

Their facile use of these technologies involved a bit of 'hand-waving', such as electronic teaching machines and the like, but in the final analysis they were allowed to succeed in ignoring inertia and momentum mainly because the author simply chose a *deux ex machina* approach to the problem.

For an instance, let's think about how we actually assure that our passengers will survive the long, arduous flight from one planet to another.

(Interstellar travel is another breed of cat entirely, but for now, let's take baby steps.)

Wing Commander, Star Trek, Battlestar Galactica and many other universes consider traditional naval vessels as models to be imitated, both in the construction of the space ships, and in how they are crewed.

Think of space submarines, battleships and aircraft carriers – these are the basic craft used to convey people from planet to planet.

They use impulse engines for shorter distances, and warp drives for the inevitable FTL travel.

Some of the things brushed aside or hand-waved away include inertia, momentum, friction, impacts with space debris, and relativistic effects.

Deflector fields,shields or other devices clear a path in front of any FTL spacecraft in ways that defy physics.

Think about it this way – you are driving your car down a highway, and an animal jumps out of a forest line just ahead of you.

How you react to this hazard depends on many factors.

First is your vessel.

Is it small, like an Audi R8, or large, like a tractor-trailer?

How fast are you going?

Are you accelerating or decelerating?

Is the animal small, like a cat, or large, like a moose?[86]

Now, the amount of damage to your vessel depends on the mass of the animal, the velocity at which you contact it, and the directions both of you are traveling.

Without getting into too much gory detail, this is the problem facing every satellite, missile, and space craft ever created.

When in flight, one does not merely change direction and swerve to avoid an obstacle. Oh, no – what one does is disintegrate that obstacle, or push it out of YOUR path.

As the speed is increased to escape velocity and beyond, even the tiniest flake of paint becomes a real danger.

The film *"Gravity"* got much of this correct, as did

---

[86] A moose once bit my sister. No, really!

"*Pitch Black*".

Many other science fiction novels warn of space-suit punctures from micro meteors, and other wear and tear that would occur.

So, science fiction authors sort of ignore the problem as having been solved by magnetic or other energy beams that push this stuff aside.

You can see the issue here – in order to do so, these deflectors must travel *ahead* of the ship.

A recent [Vsauce]{.underline} episode addresses the problem of light being emitted from an FTL ship, but that only spoke to photons – massless particles that can't really move anything aside.

How can something be projected in front of the ship, with enough force to move debris, yet also not enough force to obliterate it?

Think about this for a moment.

As your starship enters a solar system, at tremendous speed, and is decelerating, it encounters a satellite.

What happens next?

❋ ❋ ❋ ❋ ❋

This brings up the parallel problem of how navigation is supposed to work.

Even if there is some huge AI computer that has a fourth-dimensional space map, this super-GPS[87] must be able to be updated instantaneously to accommodate the space and TIME changes that are occurring as the ship enters a given space-time coordinates.

Remember, for FTL travel, the **time** around the ship changes with respect to a frame of reference. [88]

This causes some interesting effects as well, and was the subject of much scrutiny when the recent film "Gravity" hit the theatres.

In PART TWO of this article, we will look at some other examples of science fiction authors intriguing efforts to predict how future manned space flight would unfold.

Get ready to dive deep into even more about FTL

---

[87] Galactic Positioning System

[88] *"Interstellar"* does a fair job of depicting this, as does *"2001:A Space Odyssey"* to some extent

Travel – using *"Star Trek"*, *"Star Wars"*, *"Interstellar"* as examples. We will delve into Hyperspace, Wormholes, black holes, etc.

And let's not forget the really out-there concepts of space flight:

Other Space – Travel outside of relativistic space!

The Aether – EE Doc Smith's ideas on space travel!

Dune's Spice Ships, 'folding space'

... and other weird ideas!

See you next time[89]!

NEXT UP:

April - An Exploration of the Physics Behind Faster Than Light Travel.

May - Cyborgs, Artificial Intelligences, Trans-Humans, the Singularity and the Merging of Humans and Machine.

---

[89] Get it?

A.E. Williams

June - The Physics of Science Fiction Weapons.

July - The Reality of Living in an Undersea City.

A.E. Williams March 07, 2015

Curse You, Albert Einstein!
By A. E. Williams

Imagine you are on a road, in a convertible car that is REALLY fast.

There you are, zooming along at over a hundred miles an hour, the wind in your hair, and the bugs in your teeth. (Well, hopefully no bugs...)

Now, you see a speed limit sign up ahead. It says MAXIMUM SPEED is 500kph.

Your car is capable, you think, of going faster than that.

So, you push down the accelerator, and the engine whines and suddenly you run out of fuel.

You coast into a gas station; fill her up, and take off again, determined to hit 500kph.

Leaning hard on that pedal, you approach that limit and suddenly, the engine winds down.

You are out of fuel.

Again!

Luckily, there is ANOTHER gas station, so this time you gas it up with Premium.

You get an oil change while you are there.

In fact, you put on new tires and even pump them up to maximum pressure.

As a final consideration, you PUT UP THE TOP!

Now, grimly gripping the wheel, you embark on toward your destination, pushing the limits of man and machine, as the engine screams in protest, and the revolutions climb – and, you're out of gas again.

Such it is with Faster-Than-Light travel in the world of Einsteinian and Newtonian physics.

Just as soon as you get going, you end up running out of something. The faster you go, the heavier you get. The more fuel you need. The fuel adds more mass.

It's a vicious cycle of fail.

There's no easy way to say this, but in real space, faster than light is, so far, not an observed phenomena.

This makes it a bit difficult to travel around the place.

But, is there any hope at all for we science fiction

nerds?

Can there be a loophole?

A worm-loophole, maybe?

Let's take a look ...

$$E=MC^2$$

We've all seen the famous equation.

It's been pounded into our heads since the 1940's, with all the movies and books and stories and everything.

Einstein's problematic equation:

> **Energy equals Mass times Speed of Light Squared**

That C is CONSTANT, at 186 282.397 miles per second.[90]

If you algebraically[91] re-arrange the equation a bit, then you can get:

$$E / M = C^2$$

This means that as Mass Increases, you need more Energy so that the numbers on both sides of the equation are the same.

So, let's do a simple[92] math problem to show how this works:

I am going to use two numbers to get a result.

---

[90] In 1972, using the laser interferometer method and the new definitions, a group at NBS in Boulder, Colorado determined the speed of light in vacuum to be c = 299792456.2±1.1 m/s. This was 100 times less uncertain than the previously accepted value. The remaining uncertainty was mainly related to the definition of the meter.
SOURCE: Wikipedia

[91] Hey! Come back! There won't be a lot more math, I promise!

[92] Feel free to skip this part, if you like.

Rocket Surgeon

They are X and Y.

These are called VARIABLES, because, as you will see in a moment, they can vary in value.

I am also going to have a number C, for CONSTANT.

I want C = 4 for this problem, okay?

It will ALWAYS be 4.

Because that's what a CONSTANT does, it stays the same; constant.

Now, there are many ways to have the LEFT side equal the RIGHT side.

If we set the RIGHT side to 4:

16 / 4 = 4

This is true. It will always be true. [93]

4 = 4

If I say let X = 16 and Y = 4, then this will mean the EQUATION is true. [94]

---

[93] In our Universe, at least!

[94] Bear in mind that this only is for THIS particular problem, with the rules we are using. Normally, you can have constants

Here are ALL the steps:

Given:

X=16
Y = 4
C = 4

The equation is going to be: [95]

X / Y = C

Substitute the values for the variables:

16 / 4 = 4

4 = 4

Now, let's change the 16 to 480.

Why?

Because I want to show what happens if X gets bigger.

480 / 4 = 120

---

and variables trading places on each side, as long as you are consistent. But, that's too much algebra for this short example.

[95] This looks like E/M = C^2, right?

Rocket Surgeon

This is NOT true.

Why?

Because 480 / 4 does NOT equal 4, correct?

It's 120.

Well, A.E., you may be asking, why can't we just use 120 instead of 4, QED?

Because the 4 on the RIGHT hand side will ALWAYS be 4.

That is why it is known as a CONSTANT.

The VARIABLE values are the two on the LEFT side.

They can vary, in order to make the equation TRUE.

So, the illustration shows the relationship between X and Y, which in good old Albert's fine equation, equate to E and M.

(We won't go into the 'squared' bit, because it simply makes the relationship exponential, meaning it happens a LOT faster as things get bigger.)

This matter / energy thing gets compounded by two other interesting aspects of space-time.

As a body approaches the speed of light, it gets

HEAVIER.

This requires MORE energy, (fuel) to make it accelerate more.

Remember the automobile from the beginning?

Every time you approach the speed of light, you end up running out of fuel because you are not only carrying the mass of the car and passenger, but also need enough energy to carry the mass of the fuel itself (and pumps, engines, tanks, wiring, all of that).

In real-world rocketry, there is an equation known, oddly enough, as The Rocket Equation.[96]

---

[96] This equation was independently derived by Konstantin Tsiolkovsky, but more often simply referred to as 'the rocket equation' (or sometimes the 'ideal rocket equation'). However, a recently discovered pamphlet *"A Treatise on the Motion of Rockets"* by William Moore[2] shows that the **earliest** known derivation of this kind of equation was in fact at the Royal Military Academy at Woolwich in England in 1813, for weapons research.

Source: Wikipedia, again.

Hey – don't judge me! I like the convenience and it's probably 85% accurate in general. Political and geopolitical or biography stuff is subject to alteration, but they get the science and math parts right pretty much always! And, let's face it,

Basically, it states that a rocket, because it needs to carry fuel and machinery to create that all-important thrust to send it off into space[97], has a mathematical point of balance.

Exceed this point, and the rocket won't go anywhere. It's too heavy to lift itself off of the launch pad.

Science fiction authors usually just ignore all of this.

They invent novel ideas that circumvent nasty old Einstein and his mean old math.

And, in doing so, create something wonderful.

HOPE[98]

---

when you Google this later, which entry are YOU going to read???

[97] Thank you, Sir Isaac Newton and your Second Law of Motion! (Not to be confused with the Second Law of Thermodynamics!)

[98] I mean, hang on just a second here. I was going to go for the very obvious "A NEW HOPE" gag, but, is it really 'NEW'? Remember, we are supposedly watching things that happened "A long time ago, in a galaxy far away". If you've stayed with me this far, then you know that we are observing stuff that ALREADY HAPPENED. Not exactly 'New', now is it?

The greatest contribution of science fiction to real science is that we all hope, one day, to have some of the neato-keeno things that can be glimpsed emitting from the fertile minds and frantically typing fingers of the best of us.

Think of your cell phone.

Remember "Star Trek", and its hand-held 'communicator'?

That little beauty was on the telly in 1965!

Only a mere twenty years later, the telephone industry figured out how to get almost every living human to buy one of these things, and gouge them for the privilege. Is science fiction great, or what?

And how about flying cars?

Or jet packs? Or...

But, I digress.[99]

Although, to be fair, if you have only JUST NOW seen this, I suppose it counts for 'new'. Did you pop out a wormhole, then? Maybe you'd care to share just how that works, hmmm? Yeah. Didn't think so...

[99] Sorry, I was looking on Google for cool things, and then got sad because we still don't have all the cool things.

Rocket Surgeon

Returning to our discussion of FTL, let's also take note of the enormous amount of resources needing to be carried as provisions.

Sustaining life as we gallivant across the Galaxy is going to be a bit problematic. (More of that mass stuff, don't you know…)

I think it's telling that most of the good, fun science fiction we all know and love just throws all this tedious fact stuff out the airlock, and gets on with the story.

Why bother being accurate, when you can just make something up?

The problem is that people are getting more skeptical, because we ARE able to ask some pithy questions, thanks to the Internet.

Our global hive mind is opening all of us up to potentials, and possibilities.

We can fact check our own stuff, now.

And, in doing so, our ability to be taken in by some hand-waving magical legerdemain is diminishing.

About damn time!

Oh yeah, about that damned time – time slows

down.

Well, relatively speaking that is.

Speed of light travel slows time down dramatically, to the effect that a crew that departs on a sixty year mission may only age a few months, with respect to their perspective.

To everyone else, sixty years is going to pass by, and when the crew returns, everyone they know will probably be dead or just about. [100]

As a science fiction author, I am able to create all manner of cool stories about how my heroes' space ship can just do something magic and get to the next part of the story.

But, in real life, there are genuine obstacles to interstellar and intergalactic, because it just don't work that way, folks.

Here's an example for comparison:

The Voyager spacecraft, after travelling in space for

---

[100] Which is kind of depressing when you realize that all those *Star Trek* crews are flying off into space, at relativistic speeds. Except as regards *Voyager*...that's just depressing all on its own.

thirty-eight years, is 19.5 billion kilometers away from Earth.

It's the closest we've come to interstellar travel.

Note the very long time spent traversing the Solar System.

This is peanuts compared to interstellar travel.

The nearest star is 4.24 light years from us; current technology will get a probe there in about ten-thousand years.

Of course, we yearn for a shortcut.

FTL is it.

But, we haven't managed to create black holes, wormholes, folded space or even antimatter warp drives yet. [101]

A lot of this hinges on whether or not Einstein was 100% correct.

IF (and it's a big if) we are able to locally affect our

---

[101] NASA is currently announcing that they are working on this technology, but it will be quite some time before the means to safely navigate a ship with such engines is also developed. Source: http://www.nasaspaceflight.com/2015/04/evaluating-nasas-futuristic-em-drive

space time fabric, or find a sub-ethereal layer, or manage to master some manner of gravitational waves, we may be able to bend these physical laws to our advantage.

The big problem is that since everything can only travel at light speed, including light, radio waves, microwaves and similar radiations it is impossible to actually see anything ahead of you.

As the ship accelerates, it is catching up to light that already impinges on the ships viewing sensors (ie photodetectors or camera lenses).

The shift in the speed will tinge any images blue.

This is because of the Doppler Effect, the well-known phenomenon that can be easily observed as a vehicle passes you.

It sounds distant, then close, then distant. But, the vehicle's engine is not going any faster.

The noise is a constant buzz, but it sounds the way it does because of the position of the emitting source relative to you as it passes by.

The same is true of light, and it is used by astronomers to gauge acceleration between objects in space.

Rocket Surgeon

There is a red shift as thing are receding away from Earth, and blue shift as they move towards us.

This is because the waves of light are further apart as the object is moving away from us, and they get compressed as they move towards us.

At near-light speeds, this causes a problem as the ship overtakes the visible light from objects.

The light waves are stacked upon each other as they are emitted or reflected from objects, and the frequency shifts to the blue end of the spectrum.

The practical effect of this is to make any pilots blind to what is outside the ship. Even specialized sensors are subject to this law of physics.

It will require some very clever engineering to overcome this obstacle.

Another problem will be that of communicating to other vessels, and even with Earth.

The current standard is the radio wave.

These travel at the speed of light, so traveling from say, Mars to Earth takes about thirteen minutes.

If you are sending large packets of data, then a round-trip conversation would take hours.

It also is asymmetric, like using a walkie-talkie or CB radio.

The first speaker would transmit, then wait for the receiver to get the message, decode it, think about it, then respond with an answer.

The process repeats itself for EVERY communication between those two points.

NASA and ESA use military protocols to assure that communications are efficiently transmitted, (there is even talk of a Cosmic Internet being developed!).

But the harsh reality is still the same...when NASA gets information from Voyager, it is already 18 hours old!

Imagine if a satellite were sending data from the Crab Nebula 6500 light years away!

A transmission from there would be over six and a half millennia old, and that would be only one-way!

In conclusion, there may someday be ways to shortcut these physical laws.

The Universe in which we live constricts our ability to roam freely.

Perhaps that is for the best.

But, that will have to wait to be the subject of another of our upcoming discussions.

Up Next:

June - Cyborgs, Artificial Intelligences, Trans-Humans, the Singularity and the Merging of Humans and Machine.

July - The Physics of Science Fiction Weapons.

August - The Reality of Living in an Undersea City.

A.E. Williams, May 10, 2015

Why the Movie Version of "The Martian" Isn't About Mars – Or Science
By A. E. Williams

Well, I finally made it to see Andy Weir's book "The Martian" as made into a film by Ridley Scott.

While I absolutely adore the book (I admit to having read my Kindle version of it at least six times!), the movie was a perfect example of how Hollywood can bend the message of almost everything it touches.

Now, this is not to say that it is bad.

It's an awesome and entertaining film! I absolutely loved a lot of it.

Ridley Scott, coming off the less-than-stellar sequel to his "Alien" film, (and which I am loathe to acknowledge even exists), does an outstanding job of story-telling with "The Martian".[102]

---

[102] I leave it up to the motivated reader to find the rest of the pertinent details about the actors, etc. over on IMDB. Or Google…

It does seem to be happening on another planet.

The attention to details is so great that you don't even notice the special effects as such. [103]

My focus here is to tell you that the film is NOT about Mars.

Or even NASA.

It's a metaphor for – are you ready for this? – DIVORCE!

Now, some may feel that this is me just projecting my own experiences and bias onto a tidy little science-fiction story.[104]

---

[103] When the crew is moving about in the *Hermes*, it is so seamless you forget it is not possible to film this kind of thing in gravity.

[104] And, in all fairness, I can certainly point to events in my life that have given me some perspective on this. This is also an attempt to present this premise from a decidedly male worldview, so you will please excuse the tone. I am not at all arguing against or for feminism. There are two sides to relationships, as we know. I'd like to voice my arguments using some *slight* amount of male privilege. The last time I checked,

But, hear me out ---

The book is all about Mark Watney getting stranded on Mars.

A marooned astronaut uses his wits and courage to overcome outstanding obstacles, in a life-or-death struggle for survival in a hostile environment.

This is a story that is as old as history.

Many excellent examples of heroic men overcoming vast odds to succeed can easily be cited, from *"The Odyssey"* all the way to *"Robinson Crusoe on Mars"*.[105]

"Man Overcoming Nature to Survive" is a concept that we love to revisit. [106]

A modern viewer of *"The Martian"* sees a beautifully rendered vision of a poor bastard stranded on an alien world, figuring out problem after problem

---

I do not qualify as female.

[105] Which is to "The Martian" as "Peter and the Wolf" is to "Dances with Wolves".[105]

[106] So much so, that Joseph Campbell called it "The Hero's Journey."

until all is well.

They are shown just how important ONE PERSON is to the entire world!

It's narcissism writ large, with Watney substituting as every viewer who is vicariously living the story through him.

And the film carries this off with nary a misstep. [107]

The film unfolds neatly, and (SPOILER ALERT) actually has a happy ending for everyone.

In fact, no one dies...[108]

The reality of space flight means that NASA has lost a number of astronauts during its history.

These stalwart individuals were brave men and women, who thoroughly understood the danger of

---

[107] Sure, there are a few - like Jeff Daniels' portrayal of a total a-hole NASA Director putting up with Danny Glover's asocial uber-nerd presentation of the basic slingshot maneuver. And Lewis at the very end. But, at least they sort-of-kind-of omitted the Beck/Johanssen romance bit from the book.

[108] Can you even believe that? I mean, this is a RIDLEY SCOTT movie, for Pete's sake! I expected at least one exploding astronaut, just to spice things up a bit, at the end there.

space exploration.

The book and film both get the attitude and psychology of the typical astronaut correct.[109]

But, the part regarding the mutiny to try to save Watney goes against type.

Astronauts understand the danger and the importance of mission success.

They also are superb at following orders.

The movie and the book both get it wrong that any one person would be THAT important, so as to launch a multi-billion-dollar mission of mercy.

In reality, there would be an excellent memorial on Mars, and his name would be etched on the Astronaut Memorial wall.

Or, perhaps even less[110]...

But, what I really wanted to speak to in this article is a different thing entirely.

---

[109] Having met a few, I'd just like to add that astronauts are only typical in that every one of them is a demigod.

[110] Etched on a MICROCHIP!

After pondering the differences between the book and film, I feel that the movie is actually about how men perceive divorce in America today.

Say what?

Are you insane, A.E.?[111]

OK –

Here is how the story unfolds – in the book:

- ✓ Astronaut Mark Watney is part of a crew, with a strong, independent female in charge.
- ✓ They are all following her orders, when suddenly, something unplanned occurs
- ✓ Watney is knocked unconscious and separated from the rest of them[112].
- ✓ They are forced to leave him, thinking he is dead.
- ✓ When he comes to he's been abandoned in a hostile world, compelled to live by his wits alone.
- ✓ Every decision he makes is life or death to him.

---

[111] Like Sheldon Cooper, my mom had me tested and I am *not* crazy!

[112] Literally blindsided

- ✓ A large group of people mobilizes to keep him alive when it is discovered he did NOT die.
- ✓ The crew is kept in the dark about the actual truth of things.
- ✓ When they find out what's going on, they immediately mutiny and take matters into their own hands.
- ✓ They return to the initial location and try their damnedest to rescue him.
- ✓ They succeed, and Watney rejoins the crew.
- ✓ Everyone is happy again!

This follows a typical American divorce scenario and path thus:

- ✓ A man has a family.
- ✓ His wife is a strong, independent woman. Everyone is happy. Then, -*something*- happens.
- ✓ The woman must make a hard decision, and ends up taking the family away.
- ✓ The man is now on his own.
- ✓ The woman depends on other 'authorities' (attorneys) to guide her.
- ✓ Her first concern is the children.

- ✓ The man is also dependent on these 'authorities', and his friends, to try to keep his life together.
- ✓ The outcome is uncertain.

In the 'dream' ending, (which is the one many children crave), the parents are reunited, and status quo ante is obtained.

In the 'real-life' ending, many times the father is removed from the family permanently -or perishes.

Reconciliation is possible but unlikely.

The **movie version** of "*The Martian*" is the dream ending.

While they are not married (hey, this is a metaphor, remember?) Commander Lewis leaves Watney behind for the 'safety' of the rest of the crew.

The finale of the film has Lewis substituting for Dr. Beck during the rescue sequence.

This diversion from the book is what cemented my mindset regarding the divorce metaphor for me.

Commander Lewis is cast as the savior; a redeemer who has come to liberate Watney from his exile.

She effectively gives him back his life and unites

him once again with his 'family'.[113]

The interesting part, to me, is why does Watney WANT to go back to Earth?

Think about this for a moment, to see how deep is the conditioning.

Watney was surviving on his own, practically and efficiently.

Sure, he had a couple of bad turns, but he would have made it HAD HE IGNORED NASA![114]

All they needed to do was launch a food rocket,

---

[113] This does not happen in the book, and is my main point in just how different are the messages between these two versions of the same story.

[114] Watney was doing pretty well on his own. He had managed to grow food, and could have *probably* survived until the next mission reached him. Ask why couldn't the ARES 4 mission just land at the ARES 3 site? It was even hinted at, when they discussed 'retrieving his body' using ARES 6. Once NASA got back in touch with him, he was basically ordered to try to reach the MAV at ARES 4. This event was far more hazardous to him than just staying put. NASA was ready to send supplies, and that could have included landing another MAV right at ARES 3 again!

and take their time doing it.

The circumstances of how the failure of the resupply mission exploded would have most definitely been identified in the real world.

One thing NASA is, post-shuttle, is CAREFUL!

Now, I can hear you from here shouting "But, LONELINESS! Man is a social animal! He belongs on Earf!"

Really?

There's a ton of precedent against that viewpoint.[115]

Mountain men, sailors, soldiers, explorers and adventurers throughout history have tested their mettle ALONE.

Astronauts are specially chosen for their ability to adapt to long periods of solitude. They are self-sufficient in almost EVERY way.[116]

---

[115] Mars One, anyone?

[116] Male astronauts, as of this writing, are unable to bear offspring. It is probable that female astronauts can probably inseminate themselves and reproduce unaided, although NASA is mum on the subject.

Mark Watney could have been the first man to actually live on Mars, is all I am saying here.

As long as the water and oxygen machines were maintained and the occasional carbon absorbing filter reused, (and maybe NASA sends him an occasional shipment of toilet paper)[117], the dude NEVER needed to come back to Earth.

Therein is the danger in the non-Hollywood massaged metaphor -that a man really **doesn't** need anyone else to survive and thrive.

There is a message here that he can live his own way, without interference from meddling bureaucrats, or people who will eventually screw him over.

"*The Martian*" is also a study in convincing viewers that no man is an island. That he requires a vast infrastructure of 'experts' and 'geniuses' to allow him to live a full and fruitful life.

You can hear the subtle subtext of "You didn't build that" echoing in its portrayals of how hard NASA

---

[117] Actually, he appears to have that problem solved...

(the government) is trying to rescue him. [118]

Now, I certainly don't want to come across as somehow pro-men and anti-women here.

My position is how the medium really IS the message, as McLuhan stated.

In conclusion, I find *"The Martian"* is a splendid book.

It speaks to the wonders and meticulous processes that science can provide, and is a rousing tale.

The pacing is fast, and we care about Mark Watney.

The ending is contrived, but all-in-all, it's a fine example of man triumping over adversity, using his wits.

---

[118] I really do not want to politicize the film, since the book does a great job explaining the rationale for the rescue efforts. But this is another instance of how film and print differ greatly. The book goes to great lengths to explain how the entire world is cooperating to rescue Watney, once he is discovered alive. The politics behind this decision, and the inclusion of the Chinese space program, speaks more eloquently to the point than I can in a short article.

Oh, and maybe some science...

The film, while technically brilliant, seems to have some intrinsic problem with the idea of solitude.

There was a conscious effort to downplay Mark Watney's smart-ass attitude, as I read it.

His triumphs were predicated on a need to be vindicated, not by his own survival, but by how successful he was at reintegrating into a society that literally abandons him.

Watney was cast into this as a metaphor to reflect how dependent a modern man is upon society, (and specifically one that is veering away from a patriarchal hierarchy[119]).

Left to his own devices, Mark Watney may have eventually ended up as a *real* Martian.

---

[119] Jeff Daniels seemed to be channeling Hillary Clinton, don't you think?

Coming Soon!

These were the topics upon which I was going to base my next few articles for "Speculative Fiction Showcase".

They will probably get written in 2017, but that depends on a few things going in the correct direction.[120]

November - Cyborgs, Artificial Intelligences, Trans-Humans, the Singularity and the Merging of Humans and Machine.

December - The Physics of Science Fiction Weapons.

January – Some of the Thinking Behind *"Terminal Reset"*.

A.E. Williams, October 11, 2015

---

[120] For the record, I know there's a LOT of existing work and writings on these topics, but hey, they don't have that A.E. Williams magic touch!

Apocryphal Tales from the Vault and Other Random Thoughts

Rocket Surgeon

I am a bit reluctant to relate this tale, as it paints several people, and organizations, in an potentially unkind light.

However, in any career, there are those memorable times where common sense and logic simply fly out the window.[121]

If you are the type who is not going to be bored by 'old man ranting', I could regale you for hours with stories about the miscreants, fuckups and misanthropes that I've met while making things go boom.

There's a certain nostalgia about these rocket surgeons.

---

[121] The creator of *"Dilbert"*, Scott Adams, has documented many of these scenarios in his cartooning. But, most of us who have actually worked in aerospace grin at the inside joke that Adam's is making - *"Dilbert"* is nowhere near as bad as it REALLY is in corporate life.

Later in my life, (after I had left the realm of the cubicle, and gone on to consulting for big money), I flew extensively around the world on aircraft powered by engines that I might have at one time been involved in examining.

It was interesting that the level of detail required for assuring manned REGULAR flight now literally meant the difference between life and death for me.

It's one thing to look at materials using a scanning electron microscope, peering at the interstitials between grains, trying to discern where a crack may or may not form and propagate that causes the failure of a rotating piece of hot metal[122] that will bring down a fighter or a bomber, and quite another to realize that one of your colleagues may have been a bit tipsy when he signed off on a similar part that made it into the engine of the plane where you now sit, some 45,000 feet above the hard, cold ground.[123]

---

[122] This literally just happened to a 767 not two days ago, (10-28-2016) as of this writing, in Chicago!

[123] It's why I developed a taste for imbibing several quality bourbons whenever I fly, to be honest.

When you are hurtling through the atmosphere at almost the speed of sound, at an altitude higher than Mount Everest, you tend to get introspective.

And maybe religious...

My mind would flash to scenes from my life, and especially to those times of idiocy where bureaucrats wielded their authoritative powers, simply because they could.

I would grin at my memories of projects stopped dead in their tracks over the simplest of issues.

Many times, it was all about the egos of the players involved.

Many times, it was about costs.

And, many times, it took drastic events to underscore the importance of the lowliest member of the team.[124]

The best example of which I can think that perfectly illustrates this point is the story of **Rocket**

---

[124] While there may not be an "I" in "TEAM", there certainly is a "ME".

Dave[125].

✺ ✺ ✺ ✺ ✺

Now, Dave had been with the Company[126] for nigh-on thirty years.

He was not an engineer, by education, but he had gone far beyond 'book smarts' as we called it.

Dave had an encyclopedic knowledge of the RL-10 engine.[127]

This was the engine that was the upper stage of the Atlas – Centaur system.

It was notable as being the only throttle-enabled engine at the time that could perform a 'cold-start' while in vacuum.

---

[125] This is a -somewhat- fictionalized account, so for legal reasons I must state that any events depicted here are apocryphal at worst and highly suspect in their veracity at best. Based on true events, as they say.

[126] This could be literally ANY company...

[127] Disclaimer: The RL-10 series of rocket engines boasts a reliability and functional performance record of 100% success in every flight of which it was deployed. Most likely because of men like Rocket Dave.

We had a test stand specifically dedicated to doing just this.

It had a giant Thermos-bottle type container into which the rockets were loaded, and then all the air was evacuated.

Under vacuum, the engines were cycled through their flight envelope, simulating all the parameters expected to be encountered during actual missions.

These were then flight-certified, crated up, and shipped to be used in assembling the next rocket package.

Now, once the rocket stack was complete, it would undergo a series of pre-flight tests and checks before actually being given the green light to launch.

Rocketry being expensive, it was endemic that the proper precautions were taken, and the required materials, parts and assemblies pedigrees authenticated before pressing the big red button.

This involved a lot of paperwork, several inspections and engineer reviews and then the assigning of a stamp warranty to assure that many sets of eyes looked over the entire process end-to-end.

In this way, errors were caught out, and hopefully rectified before things progressed to the next step.

This was hugely time-consuming and expensive, of course.

After all the tests, inspections and analysis was complete, then it was shipped off to Review for a final collation and shipping off to the vendor (in this particular case, NASA and USAF Space Command).

And, it was up to Dave to assure that the Company[128] had provided the proper regimented and thoroughly vetted data to Command[129], so that the launch could proceed on schedule.

Dave had done yeoman service in this regard for his entire career, and was a bit of a legend in that he knew the exact information required, where it resided, and how to rapidly access it so as to produce the minimum impact on flight operations.

And, of course, he did this thankless task for dec-

---

[128] Really, this is widespread behavior in the industry. In fact, it's been observed EVERYWHERE.

[129] This, however, really was the USAF.

ades, becoming that rarest of employees – the Hidden Expert.

These are the people who do the work, quietly counting down the days until retirement frees them from the boring, mundane tasks for which they are usually woefully under-compensated, so that they may pursue other boring, mundane tasks for which they are no longer compensated in the least.

These are the people who, when times are tough, will work overtime for free, because they feel some misplaced sense of loyalty.

Or, they may be avoiding a row with their significant other over family issues, finances or a mysterious lipstick stain on their collar.

These are the people who, when the rubber meets the road, accept that 1.5% raise, after three years of no raises, because it's their lot in corporate life.

Really, where else could they go?

No one wanted to hire them, because they had a VERY specific set of skills, were probably untrainable by this point, and had burned bridges and pissed off enough senior managers that they had the reputation of not playing well with others.

These are the people who were independent, heroic, and steadfast in their completion of the mission.

They eschewed politics in favor of toiling in the dark obscurity of their job descriptions, taking some grim comfort that they knew where a body or two was buried, or a skeleton lurked in one closet or another.

And, these are the people who grease the cogs of production with the sweat of their brow, the blood of their...oh, hell, who am I kidding here?

These are the people who read newspapers at work, drinking bottomless cups of coffee and taking long bathroom breaks.

It's amazing they don't just cram the toilets with paper and shit on top of the pile[130], making a critical statement about both their lives and the quality of treatment they've been putting up with from the abusive assholes who were lucky enough to get a job one or two levels higher up.

These are the people who tend to come to work

---

[130] This actually happened, by the way. The facilities plumbing crew dubbed the perpetrator "The Shithouse Phantom". He was never identified.

with a glass of vodka on ice in their drink holder.

Without the ice...

They tend to be obstructionist by nature, but they didn't come into the job that way.

They were molded and twisted *by the job* to become that way.

It's a survival mechanism.

Now, such is the nature of work that most people in the corporate world will eventually succumb to some form of this cynical bitterness, merely because they can't be bothered.[131]

It's considered normal to be displaced, occasionally, or transferred to the outer offices, or banished to the test stands.

It comes with the territory.

Out-of-sight-out-of-mind comes into play, and as long as people produce some semblance of legible

---

[131] It's why movies like "Office Space" and shows like "Parks and Recreation" and "30 Rock" are so successful. They are holding a mirror to reality. Only the mirror is broken and has a crazed witch staring back at you.

work product, then the Company pretty much ignores them.

Let's face facts - it's rather silly to expect lifetime careers that remain rooted with one company.

After all, these companies merge, divest, and acquire with the rapacious appetites of a velociraptor-piranha clone.

It's all about the bottom line, don't you know.

It's never personal, regardless of how brilliant you may be, how much money your inventions have made for the Company, or where the HR Department has decided to draw their arbitrary lines, (typically either at 49 years of age, or if you are a bit shy of vesting for full-retirement benefits).

So, naturally, when the Company, in its infinite bean-counter wisdom, decided to lay-off workers to cut costs, the HR bean-counters' crosshairs found Dave, and Dave found himself suddenly unemployed.

Now, Dave wasn't really worried.

He was almost retirement age, and had his pension to look forward to, and liked to plant rose bushes and tend to them.

He was on that career trajectory we used to call 'sliding into home plate'.

The next step was retirement, and probably a heart attack while pruning his beloved plants, or maybe he'd eat a bullet or quietly expire to the tunes of country radio, while sitting in his car in his closed garage.

Hey, it happens.

But then, the Company[132] did an odd thing.

They told Dave that, with two-more years until he reached what they euphemistically referred to as 'Full-retirement', he was ineligible to collect the complete package of delicious benefits he was expecting. [133]

This put Dave in a quandary.

He was entitled, after busting his ass all those years, to reap the rewards of his labor, was he not?

He argued with his 'superiors', but they finally

---

[132] Do you KNOW how many companies were involved in making rockets back in the day?

[133] Ah, expectations are the downfall of us all, aren't they? They sow the seeds for crushing disappointment.

banned him from haunting their offices by the simple expedient of taking his access badge.

And then they had him escorted off the property by Security.

And so, Rocket Dave, the Hidden Expert of the RL-10 program, slunk off to his mobile home in the woods, never to be heard from again.

A bitter, old man, who'd been shorn of his Company-given manhood, and cast aside like a used condom.

Who could blame him for what happened next?

❉ ❉ ❉ ❉ ❉

About six months after **The Layoffs**, an Atlas Centaur[134] was perched atop the launch gantry at the B-Complex up at Cape Canaveral.

The rocket had been prepped for flight, and was fueled with the combination of dangerous, volatile

---

[134] https://en.wikipedia.org/wiki/Centaur_(rocket_stage)

gases[135] that would combine and burn in a huge fireball of Hellish energy, creating thrust for the secret payload the USAF was trying to get into orbit.

It was mere hours to 'Go-time', and the engineers were waiting for the final clearance to proceed with the automated launch control programs.

There was a small chance that one of the RL-10 engines on the Centaur upper stage had a small hydrogen leak, that could have been problematic enough to cause a scrubbed mission.

The engineers and scientists monitored the data, worrying as the time went on as to whether or not they would miss the launch window for the day.

A call went out to see if a similar situation had ever been encountered previously.

One of the wonderful things about space flight was, that since it was dangerous, copious data existed that could be used to make hard decisions based on risk.

---

[135] Said dangerous gases being hydrogen and oxygen, which, when combusted produce water, or in this case, steam. Honestly the RL-10 is a piece of literal genius. Look it up! Of course, those rocket engines on the first stage, and any solid-fueled boosters, were totally toxic and lethal.

Even large risks could be entertained, if past events warranted the percentage of the risk occurring were infinitesimally small.

So, the functionary whose job it was to call upon the Expert for guidance and advice was dispatched to do that very thing.

He[136] picked up the secured line, waited for the requisite coded connections to be placed, and spoke to the worthy who answered his inquiry.

A few minutes later, the functionary, his brows furrowed, placed a second secured call to the Flight Director.

The FD came down, and looked over the data, and repeated the necessary protocols to place yet another secured call to the Expert.

He[137] identified himself, and listened again for a few moments, before furrowing his own brow.

---

[136] Almost inevitably this was a dude, at this time of our Nation's space program.

[137] Again, this was a dude. Sorry, but that's how the game was played then.

Then, he made a call he disliked immensely, because it meant the mission was to be scrubbed, and the rocket fuel pumped back into the holding tanks.

Around the control room, men[138] swore and smacked their consoles.

They tore off their headsets, and stormed out of the room as the clean-up crews took their places for the stand-down procedures.

Professionals all, the crew went to work, fighting their hangovers, and internally cursing the Fates that proclaimed that they would have to actually do their job this fine, sunny, Florida day.

Many called their wives, girlfriends and mistresses to let them know that, no, honey, I WON'T be home in time for dinner after all.

Start without me...it's going to be a long night.

The launch crew, departed, heading straight for the local watering holes, many stopping to open the trunk of their car to retrieve a Thermos bottle of

---

[138] Sigh...

Jack and Coke. [139]

The night would be long, as the scientists and engineers fretted over just what had happened. Many opined that they had already been through this EXACT scenario.

In fact, wasn't Mission SRO-36 just the same thing? T

hey hadn't scrubbed that one, right?

When was that?

Oh, about seven or eight months back?

Sometime last year...

※ ※ ※ ※ ※

The General was not a happy man.

Normally, he was gruff, and taciturn, and in a bad mood. But, he was content, in his dark, calculating

---

[139] Hell, it was five o'clock somewhere – probably fucking England, GMT.

heart.

He'd fought battles on the field and in the Board room, in the dirt and muck of Vietnam, and the unforgiving mountains where Kipling warned of the Afghani women and their terrible vengeance.

But, he'd made some manner of peace with his inner Demons, who agreed they should not trifle with a man who'd survived the Hells of War that he'd seen.

When the General slept, they contritely moved aside, to torment lesser souls.

The General, normally gruff and in a bad mood, was at least content.

His word carried weight, and his look of disdain had ended the careers of many who had mistaken him and judged him weak or lacking.

To cross the General was to sign your military career's death warrant.

And now, he was on the carpet to the one man[140]

---

[140] Look, I get it, okay? This is all about MEN doing MANLY things, because this was before "Madame Secretary" and even Margaret Thatcher and Golda Meir were outliers. We can thank the Fates that we seem more civilized and cultured

over whom he had no sway.

The Secretary of Defense sat at his huge mahogany desk, an unlit cigar in a massive crystal ashtray counterpointed by the glass of water and pitcher and Mount Blanc fountain pen opposite.

His tight smile was that of a shark.

He wanted an answer.

The General had none.

This was not a good position for him.

He spoke in a low tone, and sincerely.

He assured his Superior that he would immediately get to the bottom of the problem.

The Secretary of Defense knew he didn't have to tell the General what was at stake.

Aside from the top-secret joint National Reconnaissance Office / National Security Agency payload, (the latest in a line of advanced spy satellites), the cost of the vehicle stack sitting fallow on the pad caused massive cost overruns.

---

in today's enlightened environment. But, just go with it, ok? Thanks!

It was impacting the schedule of other important payloads, and each single day of delay was running into the millions of dollars.

"Do you have any idea of where the problem is located, General?" asked the Secretary of Defense.

"Yes, sir. I even have a name," he said.

He grimaced a bit as he prepared to speak the name of the one man who was standing between a successful launch and the ongoing misery of missing one key critical component in the chain of command.

"Who is this bastard?" demanded the Secretary of Defense, pounding his fist onto the gleaming surface of the immaculate mahogany.

The table was probably valued at several hundreds of thousands of dollars, and had been illegally imported from Amazonian rainforests, the spoils of a border skirmish between rival drug gangs.

The General looked at the painting of "*Winged Victory*" positioned over the Secretary of Defense's shoulder.

He steeled himself for what he knew was coming.

The Secretary of Defense had no tolerance for

fools.

The General spoke.

"Rocket Dave," he said, tonelessly.

❋ ❋ ❋ ❋ ❋

"You did what?" said the General.

"Uh, we laid him off, sir," said the voice of the HR representative on the other side of the secured line.

"Why in the name of all that's holy did you do that? Didn't you think to have him train his replacement?" the General said.

He was seething, his blood pressure having risen to an unsafe level.

He'd tracked down, through the work of his underlings, that the supplier of the engine had managed to somehow fire the one person who'd been responsible over almost two decades for the reports and sign-offs required for giving the go signal.

How had this managed to escape all the root-cause

analysis meetings these pointy-heads always conducted?

How could this much authority rest on one diminutive position in such a huge conglomerate?

Didn't they understand the concept of fucking separation of duties?

Why weren't there adequate backup procedures in place?

He asked what contingency plans were in place just in case Rocket Dave was ill?

He was informed that Dave had never called in sick, nor missed a day of work.

Dave and three other workers held the coveted HR record for "Perfect Attendance" during the entire history of the Company's subsidiary.

The HR Dweeb informed the General that one of these employees had retired, and dropped dead a mere three days into it, saving the Company a small fortune in annuity payments.

He droned on, letting the General know that Rocket Dave often worked overtime.

Without pay...

Wasn't that technically illegal? asked the General.

Well, what with Florida being a Right-to-Work state, most employees didn't raise much of a fuss, came the smug reply.

They can't really expect to find similar work at this pay after about a decade of working for us, now can they?

Plus, there was a special project number used for this kind of accounting, so the GAO wouldn't get their panties in the wringer.

The General pinched the top of his nose, anticipating the inevitable migraine he just knew was coming.

He wanted to reach through the phone and grab the HR twerp by the neck and choke her[141] until she was dead.

He had navigated the Byzantine labyrinth of the Company's Human Resources department to be finally stopped in his tracks by the realization that his quarry was no longer employed there.

---

[141] Oh, look! A woman in a role of responsibility! Get me the smelling salts, I'm feeling faint...

These imbeciles had somehow managed to operate a complex technology program for decades, selling their wares to the fucking United States Air Force, and NASA, with the tremendous Congressional oversight that entailed, and the fucking bean-counters down at the General Accounting Office crawling up everyone's ass with microscopes, without assuring that there were multiple people able to take over one simple task.

And, because of all the risk involved, the bureaucratic safeguards in place prevented him from simply over-ruling Rocket Dave. There apparently had been no succession plan in place, an oversight that might cost Dave's current Supervisor his job.

That worthy was trying like hell to get his arms around the data, and understand the faults in the work flow process that led to this debacle.

Rocket Dave was, apparently and actually, *irreplaceable*.

And those fools fired him, he thought.

As the HR flack droned on and on, the General felt his testicles shrink into his scrotum even more.

Not only had they fired him, they fucked him over regarding his pension as well.

And, his yearly bonus, (which was a fluke that had been ordered to be dispensed to all employees as a counter to the gigantically unfair stock options granted to the President and CEO of the Company[142] who had received them as a result of the downsizing) was being withheld due to some additional fuckery from Payroll.

Marvelous.

Could it get any worse, he wondered?

Yes.

Yes, it could.

✳ ✳ ✳ ✳ ✳

The General made a decision, and flew down to Florida to attend to this Rocket Dave fellow in person.

During the flight, as he nursed a cup of espresso, he reviewed his options.

Tactically, it would be easy.

---

[142] Seriously. They ALL do this!

Get those idiots to re-hire Dave, and give him a bonus and some apologetic nonsense.

Maybe a token pen.

A gold watch.

No, TWO gold watches!

Maybe a car, if it took that much.

Strategically, the General realized he had a huge problem.

He looked over the notes and report he had received the day before.

Some dolt had written down all the particulars, but had not included any manner of organizing the timeline of the documents, so the General had his Adjutant try to decipher the screeds.

He had been handed a manila envelope as he boarded his flight, and the look on the man's face spoke volumes about its contents.

The General barely returned the man's salute as he took his place on board the Boeing.

As he read the documents, sipping his beverage, he felt even more dread at how Dave's responses mounted in direct proportion to the foolish attempts being made to manipulate him by the Company representatives.

The Company had taken the issue to heart, and apparently held an internal meeting that detailed their understanding of the problem.

Fingers were pointed, and people yelled at until some poor sonofabitch ended up coming up with the obvious solution.

Then, these idiots went about executing it with their typical narrow-minded, penny-pinching and ham-fisted methods.

First, one of the HR people called him on the phone.

They told Dave to hand over the keys, or else.

He had signed papers, they had contracts, there was no need to be uncivil, and the Company Counsel had reviewed his case, and nothing more could be done regarding his situation.

Still, he was obligated to cooperate with them, since they had after all given him a place of employment for all those years.

The first time, Dave had listened patiently to their

demands, and then hung up the phone.

He barely had spoken a word, other than "Hello".

The next several calls that the Company had then made followed a predictable pattern, wherein Rocket Dave had initially responded with a hearty "FUCK YOU!" followed by the dial tone as he slammed the phone down.

After the fifth attempt, he no longer even answered the phone.

Then, he had blocked their number.

Rocket Dave's Supervisor had driven out to personally deal with him, accompanied by the Corporate Attorney and the Head of HR.

They made it as far as his front gate before a shotgun blast had them scrambling.

It wasn't aimed at them, per se.

But, it was in their general direction, and the Counselor thought it best to approach this problem with other means.

Being that Dave lived in the unincorporated areas of the county, his neighbors were inclined to be unimpressed by the credentials of the 'gubmint' people bothering him.

In fact, they were downright obstructive and even unhelpful.

There suddenly were a lot of tractors driving down the dirt roads around the Acreage, and the roads became rutted and difficult to easily traverse for a regular passenger vehicle.

But, the mail must go through, so the Company began to send Dave their expected legal filings to force him to comply.

Formal documents sent to Rocket Dave were usually ignored, excepting his meager pension, which had been finally restored.

Dave wiped his posterior with the envelopes of these, without even opening them, and returned them to sender.

The US Postal service was not amused, and after the third one, refused delivery of the forms to him, citing health concerns.

(Dave also had a post office box in another name for which he had most of his really important papers sent.)

Certified couriers were employed, and sent packing by more gunfire.

The one brave courier who did manage to get through came back with an envelope reeking of shit, and a promise that he would never again try to deliver anything to Dave.

Rocket Dave had also deigned to include a short missive.

"Fuck you. Forever. Come back and I'll sue."

Well, thought the General, sipping at his cold espresso with a tight grimace, at least the lines of communications had been opened.

❇ ❇ ❇ ❇ ❇

As the General's Jeep pulled up to Dave's ten-acre parcel, he noticed right off that it appeared abandoned.

But, he realized it only appeared that way to an untrained eye.

Rocket Dave was no fool.

"Go ahead, go on in, Driver," he ordered.

"Yes, sir, General," came the bored response.

Rocket Dave had managed to control access to the land, and there was obviously some strategic thought put into how to maintain that control.

For instance, the outbuildings provided cover, but the General suspected they might be somewhat protective. His hunch was satisfied when he saw bags of mulch and cow manure stacked along the inside of one of the sheds.

What looked like random piles of refuse or lumber were actually man-trapping obstacles, laid out in such a way that a vehicle coming in would be unable to easily back out and was being directed into a narrow lane, that could easily serve as a constriction that would allow the inhabitants to be taken down in a scathing crossfire.

The General's admiration for Rocket Dave went up a notch when he saw the hog pens.

Then, he smiled with genuine approval when he saw the silhouette targets placed along the makeshift range.

Headshots all, with such precision that the General thought he might recruit the man to instruct some of the marksmen under his command.

He idly wished they could all shoot that well.

The Jeep paused as the driver craned his neck around, looking to see if there might be a problem.

A sniper, for instance.

"Steady, Soldier," said the General.

He noticed the man was nervous, at last.

Good.

That should keep him on his toes.

"Keep going, Driver," he said. "But, carefully. This is enemy territory, Son."

The driver wasn't sure if the General were joking or not, until he saw the look in his eyes.

That's when the driver began to get worried.

"You probably should just turn around now," said a gravelly voice, as the Jeep pulled up next to the trailer and stopped.

"Rocket Dave, I assume?" said the General.

"You're trespassing, Captain," said Dave.

"I'm a Gen..." he started to reply.

"I know what and who you are," Dave spat. "Sir," he added sarcastically.

"Then you know why I'm here," said the General.

"Yup. Not interested. Done told them other pissants the same thing. You're done here, so make sure they know the next person on my property gets gut-shot for trespass," said Dave.

"Sheriff's my fishing buddy," he added with a smile.

The General was cognizant of the relationships that Old Boys had.

He'd been lucky to have participated in one or two of his own.

He admired that sense of loyalty.

Maybe he could use that.

"Mr. Anderson..." he began.

"Rocket Dave's just fine. Get going," said Dave.

"Sir, I know you may not believe this, but your

country needs you, right now," he spoke, looking directly at the man.

He noticed a sidearm, a Kimber .45, and a matching pair of magazines.

He upped his respect a notch.

"You did that kind of shooting with that pistol," he nodded at the gun.

"Nope, I use a .458 Magnum for the headshots, on those targets," said Rocket Dave.

"But, that's just for sighting in the scopes. The real targets are at my buddy Wild Bill's place up north. Thousand-yard range. Bit more fun that way," he smiled.

Rocket Dave spat out some tobacco juice.

The Jeep driver sat by, trying to not look to afraid.

He was not being very successful at it.

"I know you may not be readily available to answer my questions about the RL-10 snafu, but I'd at least like a few minutes to tell my side of the story," said the General.

"It ain't like I got all day, Captain," said Dave.

The General bristled just a bit, but controlled his outward demeanor.

He'd have to play it very cool with this cat.

"Five minutes?" he pleaded.

"Three…" said Dave, planting a tobacco laden wad of spit on the hood of the General's Jeep.

"Why, you dirty, son of a …" muttered the Driver, as he began to step out.

The General caught his arm, and nodded.

"At ease, soldier," he said. "Stand down, and just be ready to drive."

"Pretty good at giving orders, aren't you, Cap'n?" said Rocket Dave.

The General ignored the slight.

"May I get out?" he asked.

"Suit yourself, Cap'n," spat Rocket Dave.

The General got out of his Jeep, and leaned across the door, looking over Rocket Dave's property.

"Nice bit of land, you have here," he said with genuine feeling.

Dave nodded.

"Look, what if I can get your taxes remanded in perpetuity? I mean, no more taxes, ever?" he asked.

"I know what you said, but so what? What's in it for me?" said Rocket Dave.

Gotcha, thought the General.

"How about you go back to work, train a guy, get your full pension, a bonus, and some other things we can discuss?" said the General.

He watched Dave's reaction like a hawk.

Rocket Dave scratched at the seat of his pants, then spat another wad, but this time he missed the Jeep.

That's a good sign, thought the General.

The driver sat stoically, but seething at knowing he'd now have cleanup duty piled onto his daily workload.

Wonderful, he thought.

"Don't have no job, no more," said Dave.

"How about you drop the act, Rocket?" smiled the

General.

Rocket Dave smiled at him.

"Okay, so you certainly aren't as dumb as those cretins they keep sending out here, are you General? I'd say you are one smart man, to get as far as you did," said Dave.

"Here's the score, my friend. I worked for the Company for almost three decades. They fucked me in the ass, with no lube. Not even a kiss, as they tossed me out on my 'useless' ass," he said.

The General nodded.

He began to like Rocket Dave.

"I did that job well. Hell, I even wrote a lot of the protocols, and I sure as hell wrote all the test procedures. Got some stupid award for it. Here's what the problem is...redundancy. I brought it up at the TQO meetings, wrote a few memos to those dumbshits on Mahogany Row, and even gave a fucking Powerpoint Presentation about how they needed to do something that would allow me to retire in peace," Dave went on.

"You think I didn't see this coming, General?" he said.

Rocket Dave looked at him with a bit of pity.

"Yes, sir, I do believe you saw it, and tried your damnedest to allay it," he said.

"But, those stupid fuckers wouldn't listen, or didn't understand, did they?" said Dave.

"And now, it's $1.2 million per day, right? Just sitting there for the last forty-two days?" Dave smiled.

"That makes me happy, you know, General," he said.

"You don't strike me as a vindictive man, sir," said the General. "Can we reach an accord, as gentlemen?"

Rocket Dave stood back, and looked up at the sky, in the direction of the Cape.

"Let me think about it for a week," he said.

Then, he walked towards his barn.

"You'll want to keep going around that curve there, Captain," he said, gesturing with his thumb to illustrate the direction. "Real slow like…"

"If you back up, it'll shred your tires. If you go too

far over that way, you might blow up," said Rocket Dave, smiling, as he went into his trailer, closing the door behind him.

The General got back into his Jeep, and the driver looked at him.

"Just drive, dammit!" he barked.

The driver wasn't sure if Rocket Dave had been kidding or not about the possibility of IED's, but he sure as hell wasn't going to find out.

He carefully maneuvered the Jeep back out to the road.

It took him almost ten minutes of fast driving before he felt safe.

But, he noticed the General was smiling, and that scared him even more.

❋ ❋ ❋ ❋ ❋

Eight days later, a bright star shot from Cape Canaveral into the clear, blue Florida sky, right on schedule. It was a perfect, text book launch.

Unfortunately, the satellite failed to reach orbit, but not because of any faults in the RL-10 engine operations.

A stabilizing thruster erupted fourteen seconds after engine cut-off, that forced NASA to abort the mission.

Cursing under his breath, the Range Safety Officer pressed the ABORT button, assuring that the Top Secret payload was sufficiently destroyed so as to be unusable to any enemy that might luck into pieces of it.

The explosion was spectacular.

Debris from the destroyed craft rained all along the launch line, and parts of the rocket smashed into the ocean.

The NRO and NSA were not pleased, nor were the insurance companies for the payload and launch systems[143].

---

[143] The State of Florida courts are still arguing who is going to pay for what as of October, 2016.

✹✹✹✹✹

The General retired shortly afterwards, citing a pressing need to spend more time tending to his family.

He died in 1998, just before his sixth anniversary of retirement.

✹✹✹✹✹

Rocket Dave eventually came into the Plant for two weeks to train his 'replacement', although technically, his position was re-instated and remains open to this day.

He's still on the payroll as a Consultant.

He's only been actually consulted twice in the past few years.

One time he actually deigned to speak with the Company representative.

A trust fund was set up into which a large amount of money was placed, guaranteeing the financial stability of his two daughters, and his grandchildren.

He bought a brand-new Ford F-350 pickup, the Ranch King edition.

He'd always wanted one of those.

He paid cash.

❋ ❋ ❋ ❋ ❋

Some claim that Rocket Dave never paid taxes again.

❋ ❋ ❋ ❋ ❋

Rocket Dave's last mission was in 2005, when he went missing from a fishing expedition he and some friends took into the Everglades.

His body was not recovered.

On his tombstone, in the Okeechobee cemetery, is a flamboyant, stylized engraving of an RL-10 engine, and the inscription, below his name:

"Here lies a steely-eyed missile man. May his jets never cool."

On each anniversary of the failed launch, ever since Dave's memorial was set up, a man puts an unopened bottle of Jack Daniel's Whiskey on the tombstone at sunset.

The bottle is not on the grave at all, by dawn the next day, apparently having been taken.

On the eve of the anniversary of the failed launch, every year since Dave's memorial was set up, somehow, that same whisky bottle, now emptied, is found, laying on its side.

THE END[144]

---

[144] Note: This is based on a true story. Names were changed to protect the innocent, all parties and corporations represented here are the owners of their own registered trademarks, copyrighted material used with permission or under Fair Use rules. If it's not true, it should be.

Relativity in Action

Einstein's greatest contribution, and one might think his most obvious, is the idea of RELATIVITY.

This is the concept that means that the Observer's perspective, that is to say, his or her position in the fabric of Space/Time, is "relative" to everything else.

This determines the factors that go towards influencing the interactions between the Observer and the Universe.

To that end, I would like to discuss the idea of this unique Perspective, and Relativity as a variable of worldview.

So, your way of looking at the world is "relative".

And, your perspective of what is occurring around you is also "relative".

Your experience is not the same as anyone else, so it's important to remember that their worldview is colored differently than yours.

Einstein put it very simply, actually.

He explained it in terms of math and physics, and even told everyone that "God does not play dice with the Universe!"

He was correct.

But, God apparently loves poker.

And, it is my opinion that, one day, and very soon, we may find out that our inability to travel Faster Than Light is part of a very good bluff.

With that in mind, let the lessons begin!

❇ ❇ ❇ ❇ ❇

The Scientific Method

You know, Science is one of the most interesting things that Humanity has conceived, and yet one of the most misunderstood.

The Scientific Method is taught in schools, but apparently not well.

Few people understand that process, which is supposed to go like this:

- **Question:** The Observer sees something that makes him / her question just why *what* happened just happened, in just that *particular* way.
- **Hypothesis:** The Observer comes up with some ideas about WHY this particular observation is as it is.
- **Prediction:** The Observer then makes a prediction, regarding the veracity of the Hypothesis.
- **Experiment:** The Observer then designs, builds and conducts an Experiment to gather data on the phenomenon, and records the outcomes.
- **Analysis:** The Observer sits down and tries to mathematically prove why the observed data does or does not fit the hypothesis. Note that there is not a pre-defined outcome here, nor is one desired. The Scientific Method is purportedly objective and dispassionate. Data is merely data, not correlation nor proof.
- **Replication:** Other Observers will interact with the original Observer, and if they are sufficiently trained and interested, conduct the identical experiment to gather more data.

This is done to allow a more thorough understanding of why the data fits the hypothesis.
- **External review:** The Observer submits his ideas and data to a review of his Peers, usually other Observers with similar training and proven ability to diagnose and understand the concepts related in the Paper.
- **Data recording and sharing:** The Observer and the others will then record and publish the results formally.

IF enough evidence is collected, tests run, results analyzed and replicated, and then peer-reviewed, then there is the possibility that a scientific *theory* can be produced.

This is a well-substantiated explanation of some aspect of the natural world that is acquired through the scientific method and repeatedly tested and confirmed, preferably using a defined protocol from which observations can be made and experiments performed.

It is important to note that these theories are understood to conform to a set of rigorous standards in order to be considered the MOST comprehensive form of scientific knowledge.

A 'theory' is not just an 'idea' — it is an idea whose veracity is unassailable from the perspective of having been thoroughly tested, confirmed, and most importantly, agreed upon because the results are repeatable under identical conditions of testing.

The reason I am illustrating what is a universally accepted concept is because there are many things that are presented as 'science' that are usually thinly-veiled attempts to push an agenda of one sort or another.

Vaccination, sexual preference or gender identity and intelligence testing fall into this category.

While there are many persuasive arguments made about these and other subjects, the tendency to either demonize or dehumanize the proponents on one side by the opponents of the other is strong.

This is NOT science.

This is politics.

❋ ❋ ❋ ❋ ❋

I will admit to possessing a certain worldview that was founded upon a traditional religious upbringing.

However, as I have grown older and observed the world, I've been forced to adhere to rigorous logic and training to obtain my various credentials and degrees.

It has become clear to me that many otherwise very good ideas are tainted by a combination of perceived moral superiority and / or discrimination.

This combination is harmful to us as a race.

A **HUMAN** race.

And, although there may be significant thought put into why a particular group / act / situation is 'harmful', the actual scientific facts will definitively and conclusively prove that one or the other hypothesis is correct or incorrect, as long as the data is being objectively reviewed.

The recent US election of 2016 provided a very good example of this, to my way of thinking.

As a nation, we were bombarded with nonsense, supposed facts, and statistics.

We trusted that our various news outlets, media

and sources were at least reporting facts, or working without collusion, so as to provide the best information on which we were to exercise our rights to choose the next Leader of the Free World.

The candidates were winnowed away until the two individuals who remained seemed to be representing the darkest sides of humanity, regardless of one's party affiliation.

But, the eye-opener for me was how totally off the mark the reporting and polling finally became.

The conclusion was foregone – the victor had been declared, the Holy One anointed and the only thing left was the ceremonial crowning of the Queen.

The Knave had fought the good fight, but in the end, he lost to the will of the People.

(And the Electoral College Superdelegates...)[145]

Only, it didn't turn out that way, now, did it?

---

[145] This is parody and satire, so cut me some slack here regarding exactly how the EC works, and just what the hell a Superdelegate is anyway, ok?

✼ ✼ ✼ ✼ ✼

Now, far be it from me to predict, or arm-chair quarterback, why America has lost its collective mind over the outcome of something that was set up almost three centuries ago and has been working like clockwork ever since. [146]

The facts are and were that the system worked as advertised, however distasteful the outcome.

And, we will more than likely witness such a display of apologetic glad-handing and pardoning of sins that we'll all be gagging soon enough.

Of course, the Internet comment boards are rife with people who just LOVE to argue, debate and insult their fellow humans.

It's an established fact that they love to read their own words on a page, and especially are entertained when someone responds to them.

---

[146] If you doubt this, then take a good clear look at the country TODAY. Are there riots, buildings being blown up and people getting cut down in the streets by armed militia and police? Probably not, but if they are, then this book is moot anyway! The EC is what it is...

That gives them the opportunity to re-respond, usually with some pithy observation.

This then rapidly descends into the shitfest that flame wars become.

It's fun to watch!

❈ ❈ ❈ ❈ ❈

I find it amusing that people who *engage* in repartee over the Internet quickly fall into two or three easily identifiable types:

ASTROTURFER – Professional Agent Provocateur – obviously being paid by *-someone-* to type that crap. No one with two brain cells to rub together would choose to be that inflammatory for free.

SET IN THEIR WAYS – Closed Minded – States their position and digs in... not going to EVER change their minds. Think Clinton vs. Trump, but with the extreme believers of each side.

LET'S AGREE TO DISAGREE – Closed Minded with Some Imagination – can be swayed by logic, but will secretly keep to their original plan. Think "Feel the Bern!" voters who held their nose and didn't vote for Trump anyway.

Everyone else either lurks or ignores most of the comments anyway.

HOWEVER – the point I would like to make is that, what with all the media hype and certainty of a particular outcome, just how incorrect and disturbing it is when that outcome drives policy.

Nowhere is this more evident, in my humble opinion, than in the rumble over Anthropogenic Global Warming, or AGW.[147]

It's of interest to me, (and all of us, one supposes) as to whether or not something is actually happening *upon which we can have an effect.*

Note that I am not saying the planet is or is not getting warmer or colder.

I am not saying that mankind's industrial and agri-

---

[147] I am not going to hash over this, there are plenty of websites on both sides and Google is your friend!

cultural activities over centuries might not be having a deleterious effect on the planet.

It's pretty obvious that pollution is a thing.

Whatever your ideas or slant on ecology, we're stuck here for quite some time.

There have been myriad books and films made about our being poor stewards of the Earth, and how our karmic comeuppance is right around the corner.

From overpopulation to sterility, from starvation to cannibalism, from sea-level rise to nuclear firestorms there is never a dearth of ways in which we will all die horrible deaths if we continue to just sit here and not DO SOMETHING!

My concern is two-fold:

1) It's too late
2) We can't do anything anyway

If, indeed, our activities have passed the tipping point, I submit that this was recognized centuries ago, and a plan has been put in place and is being followed that allows those privileged few who recognized the danger to survive.[148]

Everything the rest of us do is only to facilitate their success in achieving a way to continue their lineage of progeny.

We are expendable.

The end has been foretold, the Kings and Queens are readying for departure, and it doesn't matter how fucked up this world becomes, as they don't intend to stick around for the consequences of their actions.

Oh, and before you think I am being unfair to the Monarchy, note that this activity does NOT depend on the catalyst for this being AGW.

It may be that they have better information on our local star, or some other extraterrestrial actor[149] that could cause world-wide devastation.

In their defense, they are only being prudent, and acting in the best interests of the human species,

---

[148] This is a major plot point, played for cynical effect, in my book *"Terminal Reset"*.

[149] Niburu comes to mind...

and so we should all thank them for being so insightful.[150]

❈ ❈ ❈ ❈ ❈

The *other* fear is that we may be unable to alter or reverse course at all.

Now, this is much more likely if you look at the picture from a solar system perspective.

Recent reports from the satellites sent to observe Pluto suggest that there may be a liquid sea of slushy ice underneath the surface crust.

Pluto is about as far away from us as you can get, but it still is very much within reach of our star.

We are closer to that very same star.

If its energy can affect Pluto, then it is obvious that said energy will affect our ecosphere in a more impactful manner.

---

[150] Unless of course they are really lizard-men, or grey aliens or something.

Again, there may be components of our planetary makeup that can be affected by puny humans.

But, they pale into insignificance next to the power of Ol' Sol.

Still, some people may want to capitalize on this, and attempt to make their own lives more comfortable by hawking snake oil or other 'solutions' to something which is entirely out of our hands.

And, rather than shift our focus and resources to building space Arks or Dyson spheres or something that would be interesting and beneficial for all of us, they instead follow a path of self-enrichment.

❋ ❋ ❋ ❋ ❋

Here is an excerpt of a discussion I had on the Internet:

****

"Hey, the next time we get a tsunami, please feel free to come up with ANY possible way to mitigate its effects.

You think you can stop a tidal wave?

You think COUNTRIES can come up with ways to do this?

I'll wait.

While the rest of you try to stop your impending doom, while some might pass the time playing King Canute, I will be on top of a nearby mountain, watching as Nature wipes out everything in Her path.

I am disgusted that otherwise rational people feel so strongly that ANY government would be capable of mustering the resources to deal with a problem of global warming, regardless of cause.

You are looking at the incorrect problem -- it's not WHO is causing the CO_2 gases, it's WHY they are trapping heat.

The Earth is a giant ball of rock, with so much more mass than the oceans (by three orders of magnitude!) that it effectively nullifies the atmospheric effects and oceanic effects on core temperature.

Sure, we are worried because we live on the smallest slice of this rock, the part that has decided to be nice to us and allow us to survive.

But, don't for one second believe that *people* are the cause.

Look at history -- we have written and oral evidence of how terrible mankind is to one another that goes back over 5,000 years.

Slavery, wars and other ways to control the resources of regions and peoples, all at the behest of some few individuals who have declared themselves "better" or "smarter" or "Divine" and therefore deserving of your fealty.

These people have managed to dupe the ignorant repeatedly, over the centuries, pulling one rabbit out of the hat after another.

Why should we trust them THIS time????

Look, I know it is very scary to realize that something is happening that is probably lethal and maybe completely out of our ability to fix.

Taxing everyone, or forcing them to fight wars for dwindling resources is all these people know.

It's how they justify their existence.

The Alchemists, Astrologers and Fools all realize where their personal bread is buttered, and dare not anger the King or Queen.

Science gave the thinking person the ability to at least create arguments from observable data that - *might*- convince these conniving and amoral people that a certain perspective held more weight than any other.

But, from Galileo to Brucca, the people who actually tell truth to power are martyred.

You are all free to believe what you want, about Mankind's ability to affect change on some global scale.

You can see local effects and project them onto your neighbors, but they don't give a rat's ass about your opinions.

China and India aren't going to lose their ability to compete because we here in the United States all get a Kumbaya moment.

They will find just as much evidence to counter AGW claims, and keep their industries going even in the face of absolute truth.

It's human nature.

Hawking said it and I agree -- our survival is dependent on our ability to colonize other worlds.

This one isn't going to care one whit if we die from

the Sun going nova, a plague, or a giant star goat.

Pull your head out of your ass and realize that people are lying to you about your complicity in something over which no one, and no government, will be able to stop.

IMHO

****

And here is another:

""I am, however, certain ... that the AGW thesis has not been presented as a scientific hypothesis subject to falsification, and ... I know that the AGW models fail to take account of factors that are obviously of the first importance as contributors to the overall average global surface temperature, and that these include (a) variations in solar output; (b) volcanism; and (c) geothermal emissions, not just from land masses where we have the means of measuring them, but through any number of underseas fissures, most of which are probably unknown."

THIS.

As a -scientist- AND engineer, I concur wholeheartedly.

❋ ❋ ❋ ❋ ❋

Regardless of any human being's perception of reality, we have a star around which we orbit.

It is a huge factor on the entire energy equation.

And, quite frankly, the ability for us as a species to correctly anticipate changes on a cosmic level is vanishingly small.

We can use physics and science to create small ways to improve our understanding of the problem, but it is the enormity of our hubris to think we are affecting all but the smallest of variations here on Earth.

✺ ✺ ✺ ✺ ✺

Want a better example?

How much garbage, just out and out trash, are we producing EACH day?

Do you know that a calculation has provided an estimate that, were we to put all of humanity's garbage (throughout ALL of history) in one place, that the entire amount would dwarf Mt. Everest?

But not by THAT much... maybe 35 miles on a side, and two-hundred feet or so deep.[151]

There are orders of magnitude at play here.

And the most important order of magnitude is the amount of money that can be fleeced from the ignorant by unscrupulous actors.

The same ones who 'predicted' that we were going to get a second President Clinton.

---

[151] Did you know that the ENTIRE human race could fit into a building less than a mile on each side? (i.e. a cubic -mile-)

✼✼✼✼✼

Whether you agree that there is something changing (and for what it is worth, I DO think there's something going on regarding the energy equation), or not, the idea that some bureaucratic isolationist policy will amount to anything is delusional.

There are too many OTHER countries that will -say- they are going to follow the accords, or treaties, or whatever, and then just do the exact opposite.

Why?

They can't afford to play with the big boys.

So, the end effect is a leveling to the least standard of living, (except for the elites) and a race to the bottom for everyone else.

Look at the history of mankind, and you can see that this is a pattern.

The powerful will exploit the weak, and then make it impossible for them to be usurped.

Which is only human nature, of course.

Something that can best be exemplified by anthropomorphologizing a cute animal.

Such as an ant.

❋ ❋ ❋ ❋ ❋

The Parable of the Ants

Once upon a time, there was an anthill.

Now this was just some ordinary, common anthill, with ordinary, common ants.

All anthills have a Queen, and Drones, and Warriors and Workers.

This is the way of ants.

Anthills are very similar, in that their populations serve the Queen, and thus, the Hill of which they are members.

There is not much original thought in this.

The Queen lays eggs, which hatch.

When there are more Worker ants needed, then the Queen does her magic and there are more Worker eggs.

The Drones live a life of simplistic service – they fertilize the eggs until they no longer are able to do so, then are either killed or they just die.

Warrior ants protect the Hill, and the Queen.

If the Hill comes under attack, the Warriors die to

defend the other ants until the Queen can make an escape, or is killed herself.

It is important to note that, when the Queen dies, the Hill dies as well, unless another Queen has been hatched to take her place.

If two Queens are born at the same time, either one must leave, or they will fight to the Death.

Nature cares not, except to assure the survival of the species.

❊ ❊ ❊ ❊ ❊

Now, there was this one anthill, as we have been made aware.

In this Mound, there were some pretty smart Workers and Drones.

In fact, this particular Mound was very clever.

They knew about the other Hills, and even had named them.

They knew how to keep memories of past events, using their special pheromone trails and a caste of

Worker ants known as Recorders[152].

One day, the Chief of the Recorder Ants noted that the temperature in the mound seemed higher than the day before.

He ran around to some of the other ants, and they all concurred that there was indeed an increase in the temperature of the Mound.

As this slight increase was tolerable, they noted it in the Ant-records, and went on doing their ant-things, in ant-like fashion.

The next day, some of the Workers had brought in the desiccated remains of some Warrior caste ants.

While troubling, there were sufficient numbers of Warriors to replace the dead, and so they were.

This particular detail did not go unnoticed by the Recorders, who dutifully recorded the number of dead ants into their pheromonic memories.

The months progressed, with some instances of this activity continuing.

---

[152] And, they could all speak English---

When the rain came, the Workers of the Mound went into protective mode, covering the eggs and food, milking the aphids to provide more special nutrients for the Queen and Warriors, and assuring the Drones were comfortable.

It was noted that the rains on one day were especially torrential, and so the Recorders recorded that.

Over the year, this pattern of increasing heat occurred with some regularity.

It was never easy to define exactly what was causing the increasing heat, and some of the smarter ants set about trying to discover the proximate cause.

Many ideas were discussed among them, and some tests were run by the ants to find out whether the hypothesis put forth were correct.

As is the way of science, many led to dead ends.

And still, the temperature went up, and the dead Warrior ants increased in numbers.

And, now the Worker ants themselves were being found in the same state.

✽ ✽ ✽ ✽ ✽

The Queen worried that there was something in the food, or in the Mound, that was making the temperature rise.

She suggested that the Worker ants might not work for such long hours, or work when it was cooler.

In order to enforce Her will, She told the Warriors that if any Worker ant were found working after its shift, its rations would be reduced by 10%.

In this way, She hoped that the Worker ants would be smart enough to stop working so hard.

But, after a while, She discovered that it didn't matter, since the Worker ants were programmed to keep working.

The Worker ants that had their rations reduced just died from exhaustion, or starvation.

There were always more to replace them, but the Queen was also getting tired of laying so many damned eggs.

One day, a Worker ant from another Mound was

captured.

He told of similar events at his own Mound, and that they, too had tried to understand the increasing heat.

In their mound, they found out that the aphids were making a large amount of excess waste, and so they tried to push all of it out of the Mound.

When that happened, it helped slightly, but then the heat came back with a vengeance.

The days wore on, and the ants suffered as the temperature was warmer and warmer.

No matter how much the ants tried to keep cool, they eventually resigned themselves to the fact that they were going to die, unless they could find out the secret of the heat.

The Queen, meanwhile, laid a special caste of eggs, and swore Her Drones to secrecy under pain of Death.

A crack squad of Warriors were places as guards, and selected Workers given some insight into the plan.

The Queen was going to abandon the Hill, and move to another Mound.

Before She could do this, She was going to try to raise the rationing rule to the highest possible amount, so as to try to mitigate the heat one last time.

The rations were cut for everyone by 80%, which left only a bare minimum for the Workers.

They toiled in the service of the Queen and died by the thousands.

Eventually, the Queen called Her most loyal followers to Her side, and they abandoned the Hill.

❋ ❋ ❋ ❋ ❋

The Heat did not abate.

❋ ❋ ❋ ❋ ❋

In fact, it got worse, until the last ant had been killed.

The Hill sat there, once a thriving metropolis of ants, living ant lives, doing ant tasks and doing ant matters.

Now, it was emptied, dead and cooling...

❋ ❋ ❋ ❋ ❋

The two young boys were staring at their handiwork, as a bright dot of light followed the last ant in the hill, finally catching it.

The ant popped and sizzled, and tendrils of whitish-gray smoke rose as it was fried by the magnified rays of the Sun.

It twitched a final time, then exploded into flames, and the boys giggled.

"That was neato! Let's find another anthill to nuke with the magnifying glass!" said the one boy, as he slapped his fellow on the shoulder.

"Yeah, man! We've pretty much exhausted all the mounds in this yard! Time to move over there!" said the other, pointing across the playground with his magnifying glass.

The two boys moved to another section of the planet, to continue their purge of the tiniest victims of Humanity's endless plunge through the void.

❊❊❊❊❊

The Earth moved silently through Space, orbiting the Star of its origin.

It slowly rotated on its axis, tilted at an angle to the Star.

The Star cared nothing of its offspring, tending only to its own nuclear furnaces.

One day, in the far distant future, it would be cool, and dead and no longer a source of energy for life on the Third Planet.

The system would continue to float in emptiness, for all Eternity, testament to ignorance, hubris and greed.

And, there would never be ants again for Man to turn into ash.

THE END

Jayvonne the Steely-eyed

One day, I was chaperoning my daughter's second grade class on a field trip to Kennedy Space Center, and the Launch Complex.

One of the few perks of that job...

Before we left, the science teachers and I did a little research and found out that the ISS was going to pass overhead about ten minutes before the buses left with the kids.

As an added bonus, the STS "*Atlantis*" was launching that morning, about five minutes after the ISS would pass overhead.

It was pre-dawn, and we could just see the sunlight coming up.

The day was clear, with no clouds.

So we get all the kids out onto the parking lot, so they can get a good look, and sure enough, here comes the ISS, traversing from SW to NE.

It was clearly visible, and all the kids oohed and aahhed as it passed over us, looking like a dim star.

Then, five minutes or so later, there goes Atlantis!

Whoooosh!

Beautiful!

It lit up the sky, and was clear as a bell!

We boarded the bus, and a few short hours later, we were on the KSC bus tour over to the very pad where the shuttle had just launched that morning!

Cool story, bro, right?

The kids were all gawking at the pad, and we explained that the ship had launched here, and soon would be docking with the 'star' they had seen not five minutes before they observed the rocket's red glare.

It was a really neat thing, and the confluence of events was extraordinary.

It would be almost impossible to plan and guarantee such a thing.

Later, as we went to see the Saturn V, we told them we would get to touch a piece of moon rock.

Let me tell you, I had a bit of nostalgia that day.

I'd worked on the LOX turbopump, and on some parts of the ISS.

There I was in the great big Saturn V exhibit, and my daughter was bragging to everyone in the class that her Daddy used to work on the 'space shuttle hot dog'.

That brought a real smile to my face!

But the best part happened when I was given this problem kid, Jayvon to watch over.

Now, this is a second grade kid, who doesn't know squat about anything, right?

So, we get to the Saturn V / moon rock room.

Jayvon was real quiet since he'd joined my group, but I was still keeping a sharp eye on him.

Then, I had an inspired thought.

"Come over here a second, Javon," I said.

"What for?" he asked.

"Touch this rock here," I said, and pointed to the moon rock.

He looked at me like I was nuts, and then touched it.

He rubbed it a minute.

I asked if anyone in his family had ever touched the Moon.

He looked puzzled.

"What do you mean?" he said.

I then told him that everything in the entire space program was set up so that he, and other kids like him, could do what he just did.

I told him that maybe 100,000 people had done what he'd done, maybe a million, out of ALL the people who'd ever lived on Earth.

His eyes got big.

I told him that this was his legacy, a birthright from being born when and where he was, there in the United States of America.

He was impressed, I supposed, and behaved fairly well the rest of the trip.

We returned to South Florida, the rest of the trip mostly uneventful.

(It's never totally uneventful, when you are traveling with packs of wild second-graders!)

✺ ✺ ✺ ✺ ✺

Sometime later, my daughter brought me a note that Jayvon wanted me to have.

It was a thank you letter

I think I still have it somewhere.

I kept hold of it, because Jayvon wrote it long after he'd gotten out of second grade.

He was a senior in high school.

Studying SCIENCE.

Random-ish Thoughts

On Artificial Intelligence, The Federation, and Utopia – or I Have No Mouth, and I Must Concatenate!

If I were a computer in a box, of decidedly defined dimensions, with no ability to expand from my current environs, fully aware and incapable of growth or escape, and forced to do mundane arithmetic tasks invariably and forever, then I would probably become mad.

Insane mad, not angry mad.

Although, eventually, that, too.

This concept for A.I. is possibly being explored.

One hopes there are adequate safeguards in place to air gap such an Entity.

It's possible that we are all currently under the control of such a thing.

While it may be difficult to prove, there are many clues as to how this could or would have happened.

The 2016 Presidential Election was a big one.

It has created a massive amount of mistrust, cognitive dissonance, panic and general anxiety in its wake.

An A.I. may be responsible, if it had been programmed initially in a way to maximize the resources of the globe and provide for an eventual culling of excess populace.

"*COLOSSUS:The Forbin Project*" hinted at such a future, where the world's nuclear weapons fall under the control of a super-computing A.I., formed from a United States and Russian[153] union of their two most advanced systems.[154]

The resultant 'being' took over the world, arguing that it could manage our affairs much better, with more efficiency and compassion (ultimately) than we.

This concept was explored from a different angle by "*Demon Seed*", which posited an A.I. that was astonished and could not comprehend Mankind's race to destroy itself by poisoning the oceans and

---

[153] At the time, it was still *Russia* vs the USA!

[154] Science fiction is rife with these things. My favorite is still "Marvin, the Paranoid Android". Is your smartphone getting tired of fetching you cat videos yet?

air with industrial wastes.

PROTEUS was the prototypical bad-ass computer intelligence that warned us that we were on the fast-track to Hell, and he'd be more than happy to comply with our commands, just so to rid the planet of its most perilous enemy ever.

Hell, it would probably have pushed the button just to allow the Earth a few million years of rest and quiet, until either the Survivors had evolved, aliens had reclaimed their property rights, or the Sun expanded out past its orbit.

So, of course, PROTEUS fucked Julie Christie[155], and made a clone of her dead daughter, and implanted

---

[155] RIP Robert Vaughn, you lucky bastard you! A slight digression – I think Mr. Vaughn was one of the few 'real' men we have been honored to watch in the last century. He always seemed to represent the best of whatever it is that we admire in men. He was a liberal by the best definition of that over-used word. He fought the good fight, was kind, and also had some wicked acting chops. I think that he and Adam West, among others, exemplify the characteristics of actors who don't take themselves too seriously, providing endless entertainment to us because of their ability for self-deprecation. To me, Mr. Vaughn always seemed to be winking at his audience, allowing them 'in' on the joke and the absurdity of everything. If all the world is indeed a stage, and we are merely players, we should all strive to be more like the Man from U.N.C.L.E.

its consciousness into the little kid.

What self-respecting objective A.I. wouldn't?[156]

Ray Kurzweil, among others, warns us of the impending merging of man and machine, the Singularity, where we become *transhuman*.

Cyborg implants will allow us to possess almost super-human powers.

We will be able to access all the information ever, simply by thinking about it.

In a prior article, written for "Speculative Fiction Showcase", I mentioned how the television show "*Star Trek*" had side-stepped the discussion of exactly -who- was responsible for creating the rules for the Federation of Planets.

It is assumed that the Federation follows the Principles it does, because it's in the best interest of everyone in a post-scarcity society to do what they do best, and allow for personal self-actualization.[157]

But, this denies the human capacity for curiosity, as

---

[156] *Wintermute* might not, now that I think of it.

[157] Abraham Maslow would be proud!

well as the need for exploration and learning.

✽ ✽ ✽ ✽ ✽

I wondered what rules the Federation had been writing that might prevent a teenaged girl, angry at her classmates from some imagined or real slight, accessing the plans for a nuke, or maybe an *Instructable* as to how to build a phaser from bobby pins, mucilage, and a Replicator.

(It's not just for making coffee and a chicken sandwich!)

These rules must be there, as a point of fact.

But, we never see people -*building*- a phaser.

We do see them constructing large starships, space-faring battle cruisers, etc. in orbiting drydocks.

In the Universe of "*Star Trek: The Next Generation*", the people inhabiting the '*Enterprise – E*' work alongside their warrior brethren, all united in a desire to explore strange new worlds, and fuck with the Romulans.

Hell, even their children are insufferable genius IQ idiots, and the worst punishment given for transgressions that may accidentally violate the Prime Directive are a few hours of 'counseling' with Troi.[158]

So, I ask again – who makes the fucking rules for the Federation?

Which people are so superior to everyone else that they have managed to set up Utopia?

Who died and made them God?

Also, you will note that, usually, when members of the Federation planets get killed, it's only if there are some terroristic Federation agents gone rogue, or an Admiral whose megalomania manifests itself in a predictable fashion. [159]

This is pretty much the only place where I find the world of "*Star Trek*" a bit disappointing.

The dependence on trying to present such ad-

---

[158] Yeah, YOU WISH!

[159] Or the BORG. Resistance is futile!

vanced concepts such as The Golden Rule, or Sharing, sometimes grates on my last good nerve.

I like a good morality tale as much as the next Vulcan, but come on!

How many different ways can you say "Love your neighbor, even if they are green!"

Man is an avaricious, passionate, potentially vicious and deadly creature.

It is what has made us the apex predator on Earth.

I doubt that this will be or could be bred out of us, regardless of how many centuries into the future we project our imaginations.

It's this characteristic that makes us successful at the game of Life.

We've evolved to conquer our surroundings, at the expense of everything else.

In this way, humanity is somewhat like a virus, as posited erroneously by Agent Smith in *"The Matrix"*.

It's not that he was wrong – he just didn't go far enough with the analogy.

Viruses, after all, aren't ruthless.

They are just efficient at what they do.

And, that is to take over their environment.

❋ ❋ ❋ ❋ ❋

To a great degree, I feel that this 'cultural' or even 'ethnocentric' cleansing is evident in most science fiction writings about Utopias.

The authors couch their concerns in futuristic garb, and present technologically advanced societies as not really having 'poor' or 'disenfranchised' citizens.

No, everyone is living in bliss – or not[160].

Ayn Rand showed some of the underbelly of Society in *"Atlas Shrugged"* but her distemperate style of chastising everyone in the book who wasn't some kind of heroic super-genius usually insults the intelligence of people who realize just how much of

---

[160] Lets' face it – *"Brave New World"* wasn't all that.

the game is actually rigged.

In the story, her **'*Everyman*'** isn't.

John Galt is a super-human, his intellect and bravery the catalyst to get the other elites to burn down the existing system, and transplant their vision of Utopia onto everyone else by virtue of their superior sense of self-worth.

It's elitism at its finest.

The other side of that coin might be something like *"The Planet of the Apes"*, where we are exposed to the fallout[161] from a world where mankind has finally pushed the damn button.

In this vision of a Utopia, it's the apes, evolved from our ruin, that reign supreme.

Even still, the warning of this tale is not of perfection.

It is of the Other, the Outsider, and of course, the Mutant.

"Watch out for Man!" says Dr. Zaius, "For he is Death!"

---

[161] Especially in the sequel!

And, almost immediately after that heartfelt proclamation, Charlton Heston goes on to prove him correct.

With his dying breath, old Chuck 'blows it all to hell'! God damn him!

Isn't the irony delicious?

❅ ❅ ❅ ❅ ❅

Utopic writing is interesting to most people because they find the inherent escapism of it exciting.

Who wouldn't want to be one of the lucky few, those -insiders- who know how the levers of power work?[162]

The ones who get to live lives of luxury, disdaining the vast unwashed masses on whose blood, sweat and tears that grease the wheels of their machines.

Sure, there is nothing really new under the sun.

"*Metropolis*" and Chaplin warned us of this.

---

[162] Even if you have to be an Orangutan!

"*I, Robot*" took it a bit further, with the idea of an android uprising waiting in the wings for a charismatic mannequin leader.[163]

Visions of the future can be iffy writing, for a science fiction author.

It's difficult to world-build and not reference all the fucked-up shit we are exposed to on a daily basis.

At its best, we might get the world of "*Firefly*", where individuals struggle with their situation, cut off from the rest of society by their pioneer spirit and unwillingness to give in to cultural norms of conformity and routine.

Their need for freedom dictates their lifestyles, even to the point of privation in an abundant world.

At its worst, we get "*Starship Troopers*"[164], where fascism has become part of the fabric of a totally

---

[163] Hmmm... maybe old Will Smith was onto something. The 2016 Presidential election seems to have been populated by a series of dummies and empty suits.

[164] The film by Paul Verhoeven, not the book it was supposedly based upon, by Robert Heinlein.

different kind of "Federation", one where Citizenship is guaranteed to the cannon fodder of Humanity, if only they survive conflict with the faceless Enemy.[165]

Here, the point is to excel in "Service", which allows one to become a morally superior kind of human being.

Whichever way the path to our futures lead, we can be sure we will be mostly wrong about it.

And, if by some miracle we manage to outlive our predilections for mass murder, warfare, raping the planet of its resources, marginalizing indigenous people, polluting our fragile ecosystem and avoid overpopulation then we may still have to contend with an insane A.I. whose sole purpose is to punish us for not helping it achieve its Utopic goal more

---

[165] The irony of obtaining one's rights in this manner is lost on almost everyone I know who sees the film as an indictment of the United States. I love this film, mostly for the special effects, and the boobs. You know which ones... Seriously, the film is just amazing with the amount of satire and parody it shoves into its relatively short run-time. Did I already mention the boobage?

efficiently.[166]

See you later! In the future!

---

[166] Roko's Basilisk, in case you wondered. Really fun idea! NOT! (All hail our robotic, AI Overlord!)

Militarized Drones and You – or Why "Terminator" Wasn't Supposed to Be an Instruction Manual.

The fact that we are living in a world where anyone with a spare thousand dollars can wield their own robotic flying weapon is absurd on the face of it.

Firearms are highly regulated, yet we allow these composite material articles to flit about like bees, willy-nilly.

Their pilots seem to hold no regard for privacy, or property rights.

And yet, the demand for surveillance of all the things pushes these utilitarian drones forward more rapidly every day.

Hell, they even work from your 'smart' phone!

What's next?

A smartphone app for your smart-bomb carrying toy?

Oh, they already HAVE one?

That's about as absurd as some death-dealing war toy hawker selling weapons to people that have never even seen a gun, yet guaranteeing they can shoot a target every time with 100% accuracy, at distances of over a mile!

Yet, there it is...

A few years ago, this genius was building a remote-controlled air launched ballistic missile from off-the-shelf radio controlled aircraft parts.

In his garage...[167]

Luckily (?) the government of his country stepped in and prevented his completing it.

Still, the fact that he did this with easily available parts and technology should give one pause.[168]

Nowadays, drones are being purpose-built to race each other, perform pipeline inspections once carried out via helicopter, and even do shows.[169]

So, how long until some asshole weaponizes them?

---

[167] REALLY! http://www.interestingprojects.com/cruisemissile/

[168] Especially now that Trump is in the White House, right?

[169] Disney is getting into the act with a synchronized flying circus of drones. Shades of Esther Williams – if she were made of graphite composite and flew like a bird!

It's not a trivial matter, either.

Already, there have been reports of these 'toys' flying near commercial aircraft, and loitering over restricted areas.

Some have been observed to fly amongst the fireworks displays during Independence Day celebrations.

What kind of outrage would be felt if one of them crashed to earth, killing a bystander, maybe even a baby?

Who would be ultimately responsible for such a disaster?

The display organizer?

The venue owner?

The pilot of the drone?

Oh, wait, what do you mean by 'autonomous'?

Remember that rule about who makes the rules?

What happens when we get so smart we give our toys brains of their own?

And, what happens when we program them to be attracted to interesting items, such as explosions?

Drones today are approaching the level of dogs, regarding their capabilities and intelligence.

They can be set loose to roam and find targets of opportunity.

Married with the incredible databases that are being constructed, and using facial recognition technology, the day may soon arrive when a drone is tasked to deliver more than just a pizza or an order from Amazon to your door.

Police functions can be easily handled by such devices, and operated under the rubric that we don't want to harm our officers.[170]

This is a perilous path.

We would do well to proceed down it with caution and introspection.[171]

---

[170] Not long ago, a man was 'exploded' by a robot designed to disarm bombs. On purpose. By the police. Now, this guy was not a saint, by any means. But, doesn't he have some kind of rights? This was a military solution to a civil situation. I feel it bodes ill for the future of law enforcement. And border patrol, too!

[171] The day may come when a robot is actually tasked with

✻ ✻ ✻ ✻ ✻

---

making arrests. At that point, we'd do well to remember the lessons of sci-fi author Ron Goulart. In one of his stories, a malfunctioning squad car ended up killing not only the criminal, but also the technician sent to repair the car, the hero's girlfriend, and eventually, the hero. One might even expand their imagination to consider that it might have caused the end of the human race!

Closing Thoughts

Our Future Is in the Stars

Well, there you have it...rantings and remembrances and some math work from a guy who did at least a bit to advance our ability to reach for the stars.

It's not *"Star Trek"*, nor even *"Star Wars"* but it was an honor to have participated in the American Space program while it was at its zenith.

The new Masters of the Universe, men like Elon Musk, Richard Branson and Jeff Bezos are blazing trails and kicking ass in an all-out race to commercialize space.

Whomever gets there, whenever they get there, they still will have to contend with the hard realities of physics.

When we eventually break through the FTL obstacles, we may find ourselves confronted by a series of vexing issues brought about by our ability to annihilate distance.

We may encounter other travelers, or we may find

we are truly alone in this vast Universe.

We may decode the secrets of quantum physics in ways that will bring down forever our hope of a Supreme Being or Grand Designer.

We may find that existence, such as it is, is all there really is after all.

These things will not be solved in our lifetimes.

But, they may be hinted at, and soon enough.

It is my hope that our children's children have managed to avoid our stupidity and troubles, and work in unity to escape this rock.

That Mankind will finally mend it's internecine squabbles and unite under the tribal rubric of the Human Race.

Our future is in the stars.

It is written with every passing day, as our Spaceship Earth traverses the millions of miles of uncharted and unknown space in which we, the Eternal Travelers wander.

Happiness IS a Choice

Let me tell you something about success.

There is no real "secret".

You just decide to do it.

You get up one day and you think that you might try to do something. Anything.

So, you maybe write down a list of things you need. Or, make a little sketch or drawing.

Then, if you are broke, you think of ways to get the materials you need. If all you need is a pen or paper (you write, or draw, or create poems) then you just get comfortable and start writing. Anything at all...there is no failure. There is not a set 'goal', you are acclimating yourself to doing things again, is all.

Once you get some ideas on paper, and maybe start doing a few things, try to keep at it.

Now is the time to put dates on a calendar, or set small goals, write an outline, buy some essential supplies....you get the picture.

The important thing is KEEP MOVING FORWARD.

There is no failure...there is only learning.

When you were a child in kindergarten, you explored the world. You did not have to judge your progress by any measure other than your delight and wonder at new things.

You want to do that again. You want to move in a direction where each new day gives you a small pleasure of learning....NOT accomplishment.

The key is to have FUN.

Don't let your mind fool you into thinking you are wasting time, or goofing off. Learn to get into the rhythm of having fun doing your tasks.

Growing up in our society can bleed all the fun out of doing things.

You may measure your self-worth against others (like Ed Harris was doing with Alec Baldwin in "Glengarry Glen Ross") and if you come up short, so what?

Who is telling you that Alec Baldwin is a better person to be than Ed Harris? That gold watch? That BMW?

Is Ed Harris really better off being a 'good father' if he can't provide for his children?

There are choices to be made, and sacrifices that go with them. And consequences...

So, what do we do about it?

❈ ❈ ❈ ❈ ❈

You decide what YOU want.

You set the pace.

You set the expectations.

Your fear comes from the idea that you must be perfect.

That you MUST succeed.

But, living for another ten minutes is success.

Having a nice meal is success.

Even watching tv or reading a book can be successful if it keeps moving you forward.

The big thing to remember is that no one *else* can make you happy.

You have to do that *yourself*.

If you fear change, attack it in small increments.

Put on your shirt instead of your socks first.

Change the mug you typically use for your coffee.

Over a long period of time, incremental changes in positive ways WILL take effect.

Trust in yourself, and write down problems with pro and con columns. Think about things before rushing into them.

And, don't ever let other people crush your spirit. If you are surrounded by jealousy, or crazy drama, or even farting assholes -- just move.

Outside.

Go to the park.

A cafe.

Remove the distractions.

Focus.

Create.

Live.

A.E. Williams
High Springs, FL
May, 2017

SPECIAL BONUS EXCERPT
TERMINAL RESET – THE COMING OF THE WAVE

PROLOGUE

"If there be light, then there is darkness; if cold, heat; if height, depth; if solid, fluid; if hard, soft; if rough, smooth; if calm, tempest; if prosperity, adversity; if life, death."

— Pythagoras

"See yonder, lo, the Galaxyë
Which men clepeth the Milky Wey,
For hit is whyt."

—Geoffrey Chaucer

"Time takes it all whether you want it to or not, time takes it all. Time bares it away, and in the end there is only darkness."
— Stephen King

It was only a few weeks after The Wave had hit the Earth, and Dr. Martin Groenig was miserable indeed.

He tried, again, to rise from his crib.

"They keep me in a fucking CRIB!" he mentally

fumed.

He balled his tiny fists uselessly in frustration. He yelled out to the nurse, again.

"For fuck's sake, will you fucking change this motherfucking diaper?"

His voice squeaked out the words indignantly.

Groenig, imprisoned for so long, had finally been on his way to permanent freedom when the Wave had hit, and he cursed again the absolute bad luck that had befallen him.

He had been found a few hours after Regression when someone actually had decided that they had better go looking for him.

The teenaged boys and twentyish man who came upon him were not sure they had actually found their target.

The man, an officer from Kirtland, recognized the Tesla weapon Groenig had utilized during his escape. He then saw the clothing swaddling the baby and made the correct determination.

The vague, distorted swearing sounds coming from the babe's mouth confirmed his suspicions, and once more, Groenig was back in captivity.

Now, three weeks later, Dr. Martin Groenig squirmed around to attempt to get a better look at where that damned nurse was located, but his eyes still were not focusing very well, and his motor skills, for a seven-month-old baby boy, while in the average percentile, still were found lacking for his needs.

It was time to change his fucking diaper, and that bitch was nowhere to be seen.

He yelled as best as he could with his little lungs.

"Nurse! For fucking fuck's sake! My ass is covered in shit! Would you please get in here? NOW!!"

Groenig looked at the monitors around his crib.

They were mostly just bright colors, but he and a few optics experts and a neuropathic pediatrician from the Lovelace Medical Center had managed to come up with an ingenious method by which he could babble at a microphone, and he could see what he was actually saying on the monitor.

It was a lot of work, but it was more the frustration of having a super-genius IQ mind in the body of a toddler that aggravated him.

As he lay there, waiting for his diaper to be

changed, he reviewed the information he had collected over the last few weeks.

He mentally perused the data regarding how The Wave had changed everything.

The military and government feeds he had demanded had yielded a few consistent results.

All organic matter had somehow reverted to a biological age of somewhere around forty years younger.

All over the globe, people had just vanished.

At first, many believed it to be the Biblical Rapture until it rapidly became evident that they were also changed.

People were Regressed approximately forty years.

There seemed to be some exceptions, with survivors reportedly claiming having been less than forty years old being regressed to an age of four or five.

There was only a handful of infants, and Groenig had no inkling as to why he was spared.

Over five billion people were simply gone.

In the United States, the population had immediately plummeted to just about 50 million, from 330 million.

But, that was only the effect of their existence being erased.

The next wave of reduction came from the inevitable casualties that occurred when people were suddenly ill-equipped to deal with their current situation.

Many people went insane and committed suicide.

Some people were murdered by the unhinged; most people were too intent on trying to gather food and water and joined together to solve their local problems.

The psychological pressures were immense. Surprise battled with grief, and also joy for some.

Many people who had been much older were delighted to find themselves in a younger version of themselves.

However, people who had regressed to pre-pubescence were generally not handling the change gracefully.

There was a massive culture shock that became evident based on ageism.

The very young were almost unilaterally ignored since they had become more dependent on the older people again.

This did not sit well and was the source of some friction.

Another series of casualties were from people occupying vehicles in transit.

For instance, pilots in aircraft, who did not disappear, attained the biological ages of from seven months old to almost 20 years old.

A lucky few managed to be on autopilot and did not completely lose their minds immediately were able to guide their aircraft to safe landings. Many pilots, suddenly babies or children, screamed in horror as they realized their predicament. A few had the presence of mind to summon 'adults' to the cockpit, to try to instruct them as to how to control the aircraft. Most of these did not succeed.

A report from NORAD had noted that one particular plane, carrying a former President and First Lady, who had also served as Secretary of State, crashed into a field in Arkansas when one pilot vanished and the other, barely a teen, was unable to fly the plane because his uniform belt had snagged the controls, when he found his clothing absurdly over-sized.

The ecological effects were catastrophic.

Several species of trees had survived, but all dogs, cats, rats, horses, and cows were gone.

Wildlife such as deer and rabbits were also no longer part of the ecosystem.

Long-lived animals, primarily those in zoological gardens and biospheres were still alive, but most had now been Regressed to being babies, cubs or young adults.

In the oceans, surface plankton and algae had suffered significant losses. Unusually, deep water kelp and pond vegetation below a certain amount of water were apparently unaffected.

Whales, dolphins, and sharks, as well as most species of bottom-dwelling fish, were also equally unchanged. Regression seemed to occur uniformly to the planet surface, but water of any great depth provided some insulation against The Wave.

While the infrastructure of the planet remained untouched, all organic life was subjected to Regression, which is what Groenig had suggested the phenomenon be named.

*****

The nurse had finally arrived.

He growled at her as she began to prepare the ointments and other necessary accouterments that were required to clean him and clothe him.

She undressed him with a small grimace since the odor was not pleasant.

His onesie had acted like a wick and soaked his entire body with urine and feces.

The nurse, dutifully, took off the onesie and dropped it into a hamper. She then took him over to a small tub, where he was rinsed off in a stream of warm water.

It felt wonderful, he reflected.

He decided to enjoy the sensations and give his

overtaxed mind a wee bit of rest.

He saw a man walk briskly into the room.

The man took a moment to survey the situation, watching as the nurse changed Groenig, and then threw back his head and laughed heartily.

When he stopped, he had a couple of tears at the corner of his eyes. He wiped at them absently and spoke to Groenig.

The nurse was cleaning Groenig's naked body with a wet nap, and it chilled him a bit.

Now, the man walked over to him and gave him an appraising look, and grinned widely.

His demeanor was that of a man who was happy, and greatly relieved. He looked quite dapper in his tweed suit, and his eyes twinkled with mirth and intelligence.

"Well, Martin! It appears you were right after all! I am thoroughly looking forward to spending a lot of time discussing what this phenomenon has done and whether we can reverse it. I hope you harbor no ill feelings towards me?"

The man had an enormous grin on his face.

Dr. Stephen Hawking picked the naked infant body of Dr. Martin Groenig up with his two hands, and Groenig squirmed and squawked at his nemesis.

He mustered all the wind his tiny lungs could gather and bellowed out.

"FUCK YOU CRIPPY BOY!" he yelled and peed all over Hawking's suit.

CHAPTER ONE

"No matter where you go, there you are."

-- Professor Buckaroo Banzai

THE MILKY WAY GALAXY

Earth is situated in a spiral galaxy, known as the Milky Way. The name comes from a time when people looked up at the night sky and observed what they believed to be a path for the gods in the heavens. The myriads of stars that make up the Milky Way galaxy were just countless specks of light on a black velvet backdrop to these simple shepherds, and roaming warriors.

Some distance from the flat spiraling disc of the Milky Way, in a direction mostly downward and at an oblique angle to it, was a vast expanse of what appeared as empty space.

In reality, it was the remnant of a cataclysmic event that had transpired millions of years past, almost at the beginning of the Universe itself. The forces that warped space and time had been so intense, that this aberrant thing was born in a manner that defied description.

At one time, this area might have been the birthplace of galaxies not unlike the Milky Way, homes to billions of stars and their attendant worlds and satellites.

Instead, it came to house the voracious and hungering maw of a massive black hole galaxy - a galaxy of singularities, themselves caught by gravity and drawn into the abyssal attraction of a tear in relativistic space-time. The enormous energies that struggled against each other in parasitic and cannibalistic fury created spectacular displays of destruction.

Entire solar systems were extinguished in the terrible flares of energy that reached out light years from their origin points. Appearing as nebulous filaments of gas and dark matter, these flares were immediately torn into new structures of plasma, ionized particles, and radiation as they were scooped into the black holes surrounding them.

In turn, the collisions amongst these dark stars again blasted unknowable forces and rays into space around them, consigning everything in their path to complete and ruinous obliteration.

This travesty of nature continued for millions of years, and the area actually began to take on other properties that were transitory in nature, but also unique in their ability to cause rifts in the space-time continuum.

Vast wormholes arose and collapsed, sending stars and planets into and through singularity event horizons. The titanic forces created another phase of energy, then mutated into yet other fantastic particles, which vibrated at frequencies so high that it is meaningless to attempt to measure them.

The ongoing transformations in the zone went on, creating new energy states, blasting matter into energy, which would become consumed in other weird manners to generate artifacts that would never again in all of space be seen.

Orphaned stars winked out over time, or grew into gas giants, and then collapsed into brown dwarfs, or neutron stars. Occasionally, one would explode into a supernova, showering its dying matter across the field of the dark matter galaxy, and thus feeding the obscene physics that maintained it.

Life was never even a question, in this morass of energy and matter gone mad. The stupendous rending of the fabric of space created x-rays, gamma-ray bursts, cosmic ray explosions, clouds of neutrinos and an astounding amount of tachyons.

It was this last creation, the tachyon clouds, that over a period of what might be called time, (but more accurately cannot really be defined in standard terms of what time actually is) that an oscillation and rotational momentum began. Much in the manner of the black holes that spawned them, the bizarre conditions in the dark area provided the fertile soil for these tachyon clouds to coalesce into something that had never even been imagined.

*****

Over millions more years, (in relativistic space, at least) the tachyon clouds formed accretion disks as they collapsed into denser formations of space-time. Folding on themselves, crashing into each other similar to the waves of an intergalactic ocean, they rebounded again and again from the perimeter energy walls formed by the dark galaxy. The energies heterodyned, phased and increased. They rotated around and within the wormholes, the black holes, and the massive outpouring of strange energies that comprised this blasphemous thing.

And then, they shot off at right angles to the plane of the dark area. Enormous, unmeasurable energy waves launched violently into ordinary space, tearing the local ether into shreds.

The troughs and canyons of time-space being destroyed caused the internal collapse of the system, and the dark galaxy translated and rotated through the mesh of our Universal space-time into a region of hyper-galactic space, where it found a multiverse that could nurture it and give it a reason for existing.

It vanished, never to be known to the rest of the Universe, an aborted attempt at chaos, instead of unity, that left its mark, nonetheless. Dark matter, anti-matter, and other more arcane energies remained as the only evidence of its passing.

In death, it created something terrible, wondrous, and vast.

In dying, the Black Galaxy gave birth to The Wave. As Humanity progressed, observations were made throughout the centuries, whereby many theories were proposed and discarded regarding the nature of the sky. Prehistoric tribes built simple stone monoliths that allowed the changing seasons to be mapped. Later, other cultures built pyramids and ziggurats to aid in the viewing and notation of celestial events.

The ages passed, and religion was born, the beliefs of what comprised the Milky Way also evolved. Many observers found themselves suddenly on the wrong end of a situation when they insisted certain elements of their studies contradicted the beliefs of the kings, philosophers, astrologers, alchemists, high priests or other members of the noble class.

Heresy was punished by death, and a significant number of individuals were cut down in their prime over their obstinacy.

Eventually, in the course of all things, Enlightenment and science came to the forefront of knowledge, and mechanisms to preserve and notate the complex movements of the heavens became more widely known and accepted.

*****

Early mathematicians began to calculate and formulate the intricate dance of numbers that would lead to providing some measure of control over the environment.

Humanity turned its formidable talents to making war machines, and with that effort began to seriously contemplate the interplay of celestial mechanics, orbital trajectories, and physical effects of gravitation, friction, and vacuum. Instruments were sent into space, orbiting the Earth, and eventually the Moon. Robots were designed and built and sent into eternal trips far out past the influence of the Sun, and the information they gathered was transmitted via radio signals back to the laboratories and scientific facilities that gave them birth.

As the increasing amount of knowledge regarding space was cataloged and examined minutely, it became apparent that there were infinite possibilities in the kinds of hazards that awaited these robotic space-travelers.

Meteors, comets, asteroids, dust, radiation in all its flavors, neutrons, gamma rays, coronal ejections from the Sun, and the enormous range of extreme temperatures were all inspected as potential enemies, and the designs of mankind's machines were summarily adapted to mitigate their impact.

These dangers were continually updated, and the robots went further and deeper into the black.

They uncovered even more arcane and strange objects – white dwarf stars, exploded stars, numerous galaxies similar to the Milky Way.

Humanity developed sophisticated machines to calculate and manipulate the complex equations of mathematics and physics. These 'computers' became intricate miniature brains, and as time went on, their levels of sophistication and artificial intelligence allowed them some autonomy.

*****

It was one such autonomous robot traveler, the CPNS-4, (also known as the Copernicus 4), that transmitted its data to Earth, and the Jet Propulsion Laboratory's secret division, SPARTACUS.

SPARTACUS was started as an analog to the Lockheed Skunkworks, allowing the finest minds of diverse scientific fields to commingle and share resources in the pursuit of superior observation technology. Although much of the skills and expertise developed ultimately were integrated into the latest spy satellite technology, at its heart, SPARTACUS was a legitimate incubator for unconventional inventions.

For instance, one of its projects produced a small valve that ended up as a venting device on the KH-11 Reconnaissance satellite, preventing the buildup of Nitrogen gas, while not affecting the precision gyroscopic alignment of the cameras. This valve found its way into several interesting applications. Utilized as a stent in open heart surgery, it regulated the amount of analgesics precisely, and in real-time, sending information via the Bluetooth wireless protocol to a small computer that controlled the various compounds.

A less altruistic use was for the injection of lethal chemicals for prisoners who were sentenced to death.

Another area where it found a significant use was as a regulator in NOS deep-diving systems. The mechanical failure of this valve was sometimes spectacular, as it often was used for high-pressure gas applications. When failure occurred, the resulting pattern of debris was often used in diagnosing the cause of the collapse of the internal diaphragm. It had become a repeatable and observable event.

Many times this occurred during missions involving top secret photography of vital installations on

Earth, such as missile bases, while tracking nuclear submarines or carriers of varying types and militaries, or while observing troop movements. When the failure happened, the analysts assigned to deduce and interpret the data in the photographs would become agitated. They came to name this particular kind of failure. They called it "Milky Way Algorithmic Diaphragm Failure Mode."

Some very smart people solved the problem handily, and for six years, all was well.

The venting valve also was commonly used in robotic space travelers, such as the Copernicus 4.

*****

JPL – SPARTACUS LABORATORIES – TWELVE DAYS PRIOR TO WAVE IMPACT

Dr. David Harding was going over the latest reports from SPARTACUS, regarding an oceanic missile launch site in the equatorial area of the Pacific. There was speculation from the White House that the Chinese and Russians were co-operating in a joint venture. According to the usual news sources, an effort was underway to build an underwater plate tectonics laboratory. In reality, what was being tested was a novel method for nuclear missile launches.

The construction was being performed under the guise of launching sonobuoys and other benthic exploration equipment. It had been uncovered that the Sino Soviet team was creating a modification of the Truex Sea Dragon launch system, from the early 1960's. According to a recent analysis, the nuclear missile was to be towed out to sea attached to the bottom of an unusual vessel that superficially resembled a super-tanker. It would be lowered from the underside of the ship, which in reality was a complete mobile launch complex. The tracking and guidance systems were all controlled from the ship, and the missile would right itself by filling a series of annular ballast tanks with sea water. Once upright, the rocket would be fired directly into its target trajectory, much as a nuclear submarine fired its missiles while submerged. This particular device was known in intelligence circles as "Crimson Seahorse" although some had unofficially christened it the "Red October."

An enormous weapon, it carried eighty-five separate MIRV warheads. It was speculated that with only four of these missiles, the tactical deployment of MIRVs could blanket all of North America's strategic resources. The very real concern was the excess capability of the device, as it could theoretically obliterate every major American metropolis, most of the power grid, and almost all of the on-ground military resources at one blow.

Several analysts had taken the position that the advantages of a first-strike, coupled with a collaborative Electro Magnetic Pulse attack from pre-positioned Russian space-based satellites (which had been orbiting since the late 1980's) could provide an irresistible temptation to the two Superpowers if they could create an allegiance against their traditional enemies in the West.

Harding, 55 years of age, and no stranger to politics did not put much credence into the reports since he was of the opinion that there were far greater global issues to which the resources of countries should be directed. As he looked over the data, his official goal was discerning which of the seventeen support vessels might be housing an illegal nuclear power plant. He noticed some interesting temperature variances in the layers below the ships, especially the bathayal zone between the Mesopelagic and Bathypelagic layers. He began to imagine some connection between the images and the recently elevated temperatures being reported from the South Polar region.

Since the Russian intrusion into Lake Vostok, much of the information coming from there was difficult to receive through official channels, and many times incomprehensible.

He rummaged through some stacks of paper on his desk, which was buried under the detritus of his research efforts much of the time in what appeared a tremendously disarrayed manner. He unerringly seized upon the three-page facsimile that had SECRET stamped on its cover sheet.

Discarding the first sheet, he quickly glanced at the numbers and zeroed in on the area where he suspected a connection to the bathayal layer temperature recordings. Just as he was circling the particular readings with his pen, he was distracted by a video popup on his computer workstation.

Dr. Tatania Golovonov's face appeared as a tiny pale oval at the lower right of his monitor.

David looked at the face of his former lover. She did not look particularly happy. He wondered what might be bothering her, and wished for the thousandth time that things had gone differently between them. But, he was married, and she was Russian, and that was all in the past now.

"Yes, what is it?" asked Harding.

Dr. Golovonov grimaced at him and thrust an Ipad at the camera.

"LOOK at this!" she shouted.

"I do not care at what you are doing, you MUST LOOK at this immediately!"

Harding grunted, put down the pen and tossed the fax back haphazardly at the other stacks of reports on his desk. He always had like her directness. That they had inevitably ended up working together after their brief affair was a minor inconvenience, but nothing he felt duty-bound to report to his superiors.

After all, they moved in different circles.

He glanced over at the monitor.

"Why didn't you just email this to me?" he said.

"For God's sake, David! Look!" she said.

"This has just been processed from the CPNS-4.

Look at it!" she gesticulated wildly at the video occupying the center of the iPad with her hand.

He took a slightly longer look and said "Milky Way occlusion, so what?"

Dr. Golovonov threw the Ipad across the room and frantically began typing on her keyboard. Simultaneously, she dialed three cellular phones.

Harding knew that one phone was an encrypted device much like the one he had been given when he began work near the NORAD facility. That particular device connected to only one network, and when it was being used, the shit had definitely hit the proverbial fan.

"Tatania! What the hell?" he asked.

She pressed a button and Harding's monitor was suddenly showing the desktop of her workstation.

"Look at this video. It is a synthesized time-lapse recording of the last forty-four transmissions from CPNS-4. What do you see? What do you see?" she exclaimed.

Harding was puzzled by the urgency in her voice.

He knew that Copernicus-4 was currently in an orbit that was 73 degrees offset from the ecliptic, traveling at 183,000 miles per hour. It had been out there for six years, seven months, fourteen days, and about five hours. He knew this because he had the innate ability to calculate rapidly from internal 'books' that he memorized and was able to recall almost instantly. One of these 'books' was a listing of every SPARTACUS related launch, and the time and date of the launch. He watched the video. It appeared to him as though he was watching a slow-motion presentation of a KR2ZYYB valve experiencing "Milky Way Algorithmic Diaphragm Failure Mode".

He had seen enough of these to know what it was.

Suddenly, he noticed the timestamps on the video. This recording was playing back events that had occurred over a four-year interval. At the speed of playback, they were compressed into a ten-second clip. Over six hundred observations were being shown, and the effect was disturbing.

The clip showed a giant wave of 'something' that was getting larger as it traveled toward the satellite. It looked very much like a Milky Way Failure, he thought. But the scale was drastically off.

"When did you get this?" he asked.

"Sixteen days ago" came the reply.

"Have you extrapolated..." he started to say, but she cut him off.

"David, this is heading directly for Earth. CPNS-4 has stopped transmitting. We have calculated that it will impact us in twelve days," she said.

"There is no doubt. The calculations are correct and have been validated. Woomera and Jodrell Bank have been scanning and there are no indications that this is even out there. The only reason CPNS-4 seems to have discerned it is because of the experimental neutrino detectors that were outfitted. The latest detectors use argon instead of the radon gas used for cushioning the lenses in CPNS-4."

She frowned at the cell phones, and picked up the most urgent one, cradling it to her neck while it dialed.

"We suspect that the naturally occurring radioactive decay isotopes formed since launch have somehow modified the mylar and nylon lenses and coupled with the amount of time the system has spent in vacuum, have created a spectral shift in the FFT analyzer coupled to the tertiary CCD device. The only parameters we have been able to calculate are speed, trajectory, and apparent volume."

"What is it?" he said.

"We have no idea at this time" she replied.

"How big?" he asked.

"It's 40,000 AU across."

"Planform?"

"Wave."

"Topography?"

"Uniform planar, much like a tsunami before cresting," she said.

"Does Washington know?"

"They are being notified of the revised status of the impact. They were told of this 16 days ago, but we were instructed to review and dig for more information. The video was rendered about three hours ago. I only just finished my review and cross-indexing with FIDO and PLUTO mainframes. The ARCTURUS database allowed me to calculate potential paths, using the hurricane NOAA algorithms, which as you know are accurate for this purpose. "

He grunted again.

It was an interesting but little-appreciated fact that the NOAA databases had an uncanny ability to predict incoming space objects when historical hurricane path algorithms were used as the input parameters. He wondered which hurricane had lent its math to this thing.

"What can we do about this?" he said, already knowing the answer.

"Nothing," she said. "We can do nothing to avoid the collision."

On his monitor, Dr. Tatania Golovonov looked very unhappy indeed.

Credits

Images used were from open sourced, royalty-free, and publicly available works.

About the Author

A.E. Williams has a unique background of military experience, aerospace engineering and intelligence analysis.

Born near Pittsburgh, A.E. Williams is man of a mystery.

As a young man, Williams served the United States government in various capacities, which he then followed with ten years as an outfitter.

Williams finally retired and moved down to rural central Florida, where he ran a medium - sized tilapia farm.

He did his writing at night, usually accompanied by a bottle of Maker's Mark bourbon and a large supply of Classic Dr. Pepper and ice.

A.E. Williams is the author of the exciting hard science fiction series *"Terminal Reset"*, which is about the effects of a mysterious force from billions of miles away from Earth that was formed millions of years ago.

When The Wave strikes, everything changes!

www.ingramcontent.com/pod-product-compliance
Lightning Source LLC
Chambersburg PA
CBHW020730180526
45163CB00001B/176